D0416905

Aviation Law & Regulation

A Framework for the Civil Aviation Industry

WA 1136318 5

Aviation
Law & Regulation

A Framework for the
Civil Aviation Industry

Carole Blackshaw

**UNIVERSITY OF GLAMORGAN
LEARNING RESOURCES CENTRE**

Pontypridd, Mid Glamorgan, CF37 1DL
Telephone: Pontypridd (01443) 482626

Books are to be returned on or before the last date below

MAY 2001	7-3-12
	30/9/12
9 JAN 2004	
23 MAR 2009	
-8 JAN 2010	
05 FEB 2010	

Pitm

Pitman Publishing
128 Long Acre, London WC2E 9AN

A Division of Longman Group UK Limited

First published in 1992
© Longman Group UK Limited 1992

British Library Cataloguing in Publication Data

A CIP Catalogue record for this book can be
obtained from the British Library.

ISBN 0 273 03446 4

All rights reserved; no part of this publication may be
reproduced, stored in a retrieval system, or transmitted in
any form or by any means, electronic, mechanical,
photocopying, recording, or otherwise without either the
prior written permission of the Publishers or a licence
permitting restricted copying issued by the Copyright
Licensing Agency, 90 Tottenham Court Road, London W1P
9HE. This book may not be lent, resold, hired out or
otherwise disposed of by way of trade in any form or
binding or cover other than that in which it is published,
without the prior consent of the Publishers.

Typeset, printed and bound in Great Britain

11363185

Contents

Part III   Liability

Part IV   The Environment

Appendices

Acknowledgements

There can be only one regret in writing this book and that must be that I did not do it some years ago. Whilst developing my career as an International Aviation Lawyer the opportunity to write more than articles and short publications never seemed to arise. However, I have now been able to produce the sort of book that would have been of great help to me during those years and I hope will in future be of benefit to others in my profession, the aviation industry and, not least, the students who I regularly lecture on Air Law both here and abroad.

Over the years many people have helped me gather the knowledge and experience which has enabled me to write this book. To all those I am grateful but I am especially indebted to those who with their particular expertise read and commented on relevant parts of the book. I would like to thank Dave Tompkins of the Civil Aviation Authority; Craig Orr and Bankim Thanki, both Barristers at Fountain Court and Katherine Holden for her scientific expertise.

I also wish to thank my colleagues at the law firm of Stewarts, where I practise as an Aviation Consultant, for their keen interest and understanding whilst I completed this book.

Finally, very special gratitude is due to my parents without whose support and encouragement this project would have been impossible.

Carole Blackshaw
London, September 1991.

Introduction

The nature of aviation law and regulation

It is perhaps something of a paradox that the apparent freedom of rapid transit through the skies is a business bound by a complex regulatory framework unique to the aviation industry. This is to some extent inevitable: the reasons being manifold and persuasive. A certain degree of regulation is clearly necessary in such key areas as public safety, operational organisation and international relations. The extent and nature of this intervention is the subject of continual debate involving many different interests both from inside and outside the industry.

Regulatory development is a constantly evolving process at all levels and this book intends to plot a clear course through the various areas of law and regulation affecting the industry and particularly the airline operator. Before enquiring further into the different types of regulation, it is useful to bear in mind the various levels at which they can occur. In general terms these can be summarised into three categories; international, national and local measures.

International level

Internationally the relationship between nations imposes rights and obligations both for the protection of individual states and to enable air transport between, over and into their boundaries. An obvious example is the negotiation and agreement of traffic rights. These Air Service Agreements (commonly known as ASAs) are mostly bilateral in form, being between two states. They may, however, be multinational agreements to which a number of nations are party. For example, the Multilateral Agreement on Commercial Rights of Non-Scheduled Air Services in Europe of 1956 regulates certain charter flights between a number of European countries. ASAs aim to enforce an acceptable balance of traffic between states. Commonly this is achieved by limitations on the number of airlines to be designated and on capacity provided on a flight or seat basis. Realistically, wider political considerations often influence the negotiation of these agreements and merely serve to illustrate the extent of governmental intervention in aviation.

International control can also arise from multinational political groupings such as the EEC (European Economic Community) or aviation organisations such as ICAO (International Civil Aviation Organisation) or IATA (International Air Transport Association), formed at government or industry level. Similarly aviation history is punctuated by a series of international conventions. These have made a significant contribution to regulation in areas of international relations, security and liability.

Notable is the Chicago Convention of 1944, which established a basic framework of regulation essential to the development of organised international aviation. The aims of the Convention were, through co-operation, to promote security and peace between nations. In particular, the representative governments met to agree certain principles 'in order that international civil aviation may be developed in a safe and orderly manner and that international air transport services may be established on a basis of equality of opportunity and operated soundly and economically' (see preamble to the Convention).

To date, over 150 nations have ratified the Chicago Convention as originally in force as of April 1947. The main agreement is in three parts, dealing with Air Navigation, the International Civil Aviation Organisation (ICAO) and International Air Transport. More specifically, it deals with such important issues as the general principles of sovereignty over air space; traffic rights in respect of scheduled and non-scheduled flights; aircraft nationality and registration; crew regulation; air navigation services and airport provision. These general principles have been fundamental to the organisation of international aviation and its continued development, particularly in respect of international rights and obligations.

The Convention begins, for example, by clarifying the fundamental issue as to each nation's sovereignty over its airspace. Article 1 states perhaps the most essential principle of all, that: 'The Contracting States recognise that every state has complete and exclusive sovereignty over the airspace above its territory'. From this follows the logical premise that: 'No State aircraft of a Contracting State shall fly over the territory of another state or land thereon without authorisation by special agreement or otherwise....' (Article 3 (c)). Much attention was paid to the subject of traffic rights, both at the Convention and subsequently, out of which has developed the system of original and negotiated rights that we have today.

It was also the Chicago Convention that created ICAO, which in many ways continues the functions of the original convention. The objects of ICAO, as set out in article 44 of the Chicago Convention, are to 'develop the principles and techniques of international air navigation and to foster the planning and development of international air transport'. Over the years, ICAO has been responsible for carrying out a review of key issues on such matters as airports, air navigation, safety and environmental issues and subsequently passing resolutions for adoption by member states. For example, it made a fundamental contribution to the review and promotion of noise regulation under Annex 16 (see Part IV on environmental issues).

Another vital area of regulation originating at the Convention level is the

Warsaw System governing the liability of commercial airlines on international flights. Over 100 nations have ratified the Warsaw Convention of 1929, the aim of which was to establish a uniform international system of legal rights and liabilities between airlines and their passengers and consignors of cargo. The problem was obvious and a solution essential. Without regulation, there would be a mass of conflicting laws arguably applicable and many circumstances where none would be decisive.

The key to the solution at Warsaw was to balance the rights of an airline with those of its passengers: in general terms the carrier airline is liable for death, injury or damage occurring during the 'carriage by air' without proof of cause or liability. In return for this there are set monetary limits on the amount of compensation payable. These limits have given uniformity and predictability but have also proved to be the Achilles heel of the system. Not only has the method of conversion into national currencies been difficult to achieve, but the standard itself difficult to agree. Attitudes towards acceptable damages in different jurisdictions vary enormously.

Subsequent conventions such as the Hague Protocol in 1955 have attempted to update and amend the Warsaw System, but to no satisfactory conclusion. The conflicting views, particularly those of the United States of America who argue for far higher limits of compensation, are stronger today than ever. It is possible, therefore, that the Warsaw system will disintegrate in future or that a system of new regulations will replace the old in this vital area of aviation law. (The Warsaw System is dealt with at some length in Part III.)

National level

Individual nations legislate to give effect to these international agreements and to control the operators of their airlines, aircraft and airports. In most jurisdictions, airline operators must comply with three basic areas of regulation: first, they must obtain the appropriate authorisations to operate an airline; secondly, every aircraft must be airworthy and subject to regular checks; thirdly, they are subject to various economic regulations and, in particular, the requirement for route licences. There are, in fact, many permits, certificates, licences and similar authorisations to be obtained before air transport services can be operated. In essence, these centre around three things: the operator, his aircraft and the routes served.

In the United Kingdom, the Air Navigation Order 1989 provides, for example, that an operator of United Kingdom registered aircraft must have in force an air operator's certificate. The Civil Aviation Authority grants operators certificates only once it is satisfied that the applicant is competent having regard to his conduct, experience and facilities. The Air Navigation Order also provides that no aircraft shall fly unless there is in force a certificate of airworthiness valid under the laws of the country in which it is registered. Further provisions deal in detail with the registration of aircraft.

These regulations illustrate the provision of detailed regulation by a particular state based on the general principles in the Chicago Convention. We will see many instances in Part II of the book (dealing with operational regulations) where the source is international, often laying down the general principles, and national legislation provides the detailed regulations. For example, the Chicago Convention provides in Articles 17 and 19 that aircraft shall have the nationality of the state in which they are registered and that registration in a contracting state shall be made in accordance with its laws and regulations. Article 31 states that every aircraft engaged in international navigation shall be provided with a certificate of airworthiness issued by the state in which it is registered.

The third basic area of regulation affecting an airline operator is that of route licensing. Such licences are valuable assets essential to the commercial viability of an airline. In the United Kingdom application is made to the Civil Aviation Authority which has a wide discretion within certain statutory criteria to grant, vary or revoke licences. Other operators can oppose applications, in which case a public hearing is held. The Civil Aviation Authority's stated policy is firmly committed to competition as being in the best interests of the industry and consumer alike. This can result in several carriers operating a particular route. This is a fundamental divergence from the one-route one-airline policy still followed by many nations. It also reflects the move away from the dominance of state-owned monopolistic airlines.

These changes, of course, reflect the movement towards deregulation led by the United States of America from 1978. Europe – and the United Kingdom in particular – is looking towards a freer and more competitive environment, where potentially licensing restrictions will be at a minimum necessary to protect safety standards and financial fitness requirements. The single market of post 1992 within the European Community is bound to have a dynamic effect on European aviation.

The final part of this book, Part IV, deals with environmental issues: a matter of increasing concern and significance to all of us. Aviation is by no means untouched by these considerations; the regulation of aircraft noise, in particular, has become an important and highly significant issue. To many airlines, it still represents a further unwelcome restriction on their commercial operations and to some a serious threat to their financial viability. This is particularly so in the case of cargo carriers and developing world airlines, where a high proportion of their fleets do not meet Chapter 2, let alone Chapter 3, requirements.

Airlines, however, ignore environmental issues at their peril. The strength of the environmental lobbies in Europe is increasing. This in turn provokes greater public awareness, criticism and eventual political action. We will see in Part IV that there are other improvements that could be made by the aviation industry to protect the environment. Whereas noise was very much the concern of the 1980s we now see aircraft emissions emerging as the environmental issue of the 1990s.

Local level

Noise regulation is also highly relevant at the local airport level, where anti-noise pressure groups are most prevalent. Airport authorities have for some time adopted local measures such as night curfews and further restrictions are likely. Of course, many other regulatory measures originate at the local airport level. Their source is normally either the Airport Authority itself or the appropriate local Government authority for that area. The latter will inevitably have been involved in planning matters and such issues as public health and local transport considerations.

Aviation regulations span many issues at all levels and provide a system of international travel that is detailed and complex but, hopefully, efficient and safe. Regulation of air transport is in a state of constant and often rapid change. The issues for the future are likely to be the legal liability of carriers with the possible disintegration of the Warsaw System, liberalisation in the European Community with increased competition in a single, common market, and, last but not least, the introduction of further noise restrictions working towards Chapter 3 compliance and no doubt, in addition the regulation of aircraft emissions.

The purpose and use of this book

This book aims to be both easy to read and informative. It is a friendly guide through a potentially vast and complex subject. It is, therefore, inevitably selective and seeks to provide a framework of information to give the reader a positive overview and general understanding of the subject. It is not intended as a legalistic tome of comprehensive proportions, but instead attempts to simplify and clarify, and is intended to be read either as a dip-into desk companion or as a broad introduction to the fascinating world of civil aviation and its regulation. For clarity and ease of reference, the book is divided into four parts, each covering a distinct and separate area.

Part I, sets the scene by taking the reader through key organisations in civil aviation, both international (with particular reference to Europe) and national, notably here in the United Kingdom. These bodies are in fact the rule-makers and, therefore, the source of much of the regulation reviewed in Parts II, III and IV of the book.

Part II, deals with the many regulatory aspects of an air transport operation. It shows the wide and extensive controls and restrictions so unique to the aviation industry.

Part III, deals with a vital aspect of air transport: the risks and dangers when things go wrong. It demonstrates the difficulties presented by international carriage on national legal systems and how the conflict of law problems has been mostly solved.

Part IV, is concerned with the main environmental problems caused by aircraft operations and how the regulators have begun to deal with these. A whole section of the book is devoted to this increasingly important subject, which affects aviation as much as many other aspects of industry and commerce.

Finally, the reader should always remember that the law is a complex, vast and ever-changing subject, especially in aviation. It is, therefore, essential that when dealing with any particular case proper professional advice should be sought.

Part I
Aviation Organisation and Control

Before commencing our journey through the labyrinth of law and regulation affecting the air transport industry, we will take a look in this, Part I of the book, at some of the institutions that dominate aviation both internationally and here in the United Kingdom. In Parts II, III and IV we will, in effect, be looking at the regulations produced by these bodies, but first we need to establish who the rule-makers are. These bodies are the source of, and authority for, the rules. We need to know who and what they are, their respective roles, functions and powers. They are numerous and varied but all have a vital part to play.

We will look first at the international bodies, with particular reference to the European Economic Community. This is followed by a review of the United Kingdom scene showing the hierarchy of national power and, in particular, explaining the important role of the Civil Aviation Authority.

1 International Organisations of Civil Aviation

The International Civil Aviation Organisation

The International Civil Aviation Organisation, or ICAO as it is commonly known, was created in 1944 as the brain child of the Chicago Convention ('Convention') of that year. As we will see, the foresight of the parties to that Convention enabled them to realise that to achieve their goals of security, co-operation and organisation in international aviation would require more than a one-off convention. They were ideals that needed a permanent organisation in which to carry on their work by establishing regulations based on the general principles in the Convention, developing, updating and enforcing them through the future years. Thus ICAO was created, and exists today as a flourishing organisation playing a vital role on the stage of international aviation.

Chapter VI of the Convention, under the heading of 'International Standards and Recommended Practices' essentially creates ICAO. It commences with the general principle that

> 'each contracting state undertakes to collaborate in securing the highest practicable degree of uniformity in regulations, standards, procedures, and organisation in relation to aircraft, personnel, airways and auxilliary services in all matters in which such uniformity will facilitate and improve air navigation'.

The key word is, of course, uniformity: the ideal is to achieve a system of uniform regulation on matters affecting international aviation. To achieve this end, the Convention created ICAO to 'adopt and amend from time to time, as may be necessary, international standards and recommended practices and procedures' dealing with specific matters.

These matters are:

(a) communications systems and air navigation aids, including ground marking;
(b) characteristics of airports and landing areas;
(c) rules of the air and air traffic control practices;
(d) licensing of operating and mechanical personnel;
(e) airworthiness of aircraft;
(f) registration and identification of aircraft;

(g) collection and exchange of meteorological information;
(h) log books;
(i) aeronautical maps and charts;
(j) customs and immigration procedures;
(k) aircraft in distress and investigation of accidents;

and finally such other matters concerned with the safety, regularity, and efficiency of air navigation as may from time to time appear appropriate.

ICAO is actually created, its objectives specified, powers stated and constitution established in Part II of the Convention headed, 'The International Civil Aviation Organisation'.

Formation

First, it states that an organisation of that name is to be formed (see Article 43). Its basic composition is to consist of an Assembly, a Council and such bodies as may be necessary. The permanent seat of the organisation is to be a place determined by the Assembly and is in fact in Montreal, Canada. (Article 45).

Status

The organisation is to be an independent body with a certain legal status defined by the Convention (Article 47). It is to 'enjoy' in the territory of each contracting state such legal capacity as may be necessary for the performance of its functions. Full juridical personality shall be granted wherever compatible with the constitution and laws of the state concerned. These theoretical rights are somewhat more difficult to clarify in reality and are subject to the usual problems of enforcement within an independent jurisdiction unless adopted as part of its law. However, the Convention stated the ideal for the future in an attempt to ensure the organisation it was creating would have the force of law behind its regulations.

Objectives

The broad objectives of ICAO are expressed as follows (Article 44): the aims and objects of the organisation are to develop the principles and techniques of international air navigation and to foster the planning and development of international air transport so as to:

(a) ensure the safe and orderly growth of international civil aviation throughout the world;

(b) encourage the arts of aircraft design and operation for peaceful purposes;
(c) encourage the development of airways, airports, and air navigation facilities for international civil aviation;
(d) meet the needs of peoples of the world for safe, regular, efficient and economical air transport;
(e) prevent economic waste caused by unreasonable competition;
(f) ensure the rights of contracting states are fully respected and that every contracting state has a fair opportunity to operate international airlines;
(g) avoid discrimination between contracting states;
(h) provide safety of flight in international air navigation;
(i) promote generally the development of all aspects of international civil aeronautics.

Constitution

There are two main bodies within the organisation (Article 48): namely, the Assembly – in which is vested the ultimate authority and control – and secondly the Council, which is in practice more like an executive body concerned with the day-to-day running of the organisation.

The Assembly

The Assembly meets annually, though extraordinary meetings can be held at the instigation of the Council or of not less than one fifth of the total number of contracting states. Importantly, all contracting states have an equal right of representation at meetings of the Assembly and each is entitled to one vote. A majority of contracting states is required to constitute a quorum and decisions are normally taken (unless otherwise provided) by the majority of votes cast. It is, therefore, a fairly simple structure: one state, one vote and (most) decisions taken on a majority of votes.

It is interesting to note that a very large and powerful state will have no more voting power than a small, insignificant one; it is fundamental to the idealism that created ICAO that it is based on democratic, non-discriminatory principles. To understand this one only need consider the objectives of the organisation as stated above.

The powers and duties of the Assembly relate to its (legislative) control of the organisation, in particular the Council. Apart from electing its own President and other officers, it elects the contracting states that are to be represented on the Council. It sets its own rules and procedure and generally controls the finances of the organisation.

It also deals with Council matters such as examining and taking action on reports of the Council and other matters referred to it from the Council, and generally dealing with matters not dealt with by the Council. It can delegate to the Council the powers and authority necessary or desirable to carry out the duties of the organisation. Amongst its other powers and functions, it can consider proposals for modification or amendment of the provisions of the Convention and, if it so approves them, recommend them to the contracting states for approval.

Thus, the Convention has become a self-perpetuating living organisation that can change and adapt itself in accordance with the needs and wishes of its members at a particular time.

The Council

The Council is a permanent body responsible to the Assembly. It is composed of contracting states elected to it by the Assembly. These elections are held every three years (Article 50). It is at the Council level − in effect the second level of the power structure − that some attention is given to the varying importance and influence of representative states. Namely, in electing the members of the Council, the Assembly must give 'adequate representation' to three categories of contracting states: first, those of chief importance in air transport; secondly, those not otherwise included that make the largest contribution to the provision of facilities for international civil air navigation; thirdly, those not otherwise included whose designation would ensure that all major geographic areas of the world are represented on the Council.

This more subjective and selective approach, taking account of the contribution of different states, is in contrast to the strictly one-state-one-vote approach of the Assembly, where each state is represented equally (see above).

In order to ensure non-biased representation of the contracting state, the Convention provides that no representative of a contracting state on the Council should be actively associated with the operation of an international air service, or financially interested in such a service. This establishes beyond doubt an essential point of the organisation that it is made up of representatives from contracting states to consider the interests of aviation generally. It is not in any sense to be an airline or trade association and does not exist to further the interests of one particular group.

The Council elects its own President and Vice President. Decisions by the Council (normally) should require approval by a majority of its members (Article 52). This normally includes the vote of any Vice President, but the President himself has no vote (Article 51). There are powers to delegate authority on particular matters to committees (Article 52) and any contracting state may participate (without a vote) in the consideration by the Council or its committees and commissions of any question which especially affects its interests (Article 53). No member of the Council is to vote in the consideration by the Council of a dispute to which it is a party.

Functions of the Council

Articles 54 and 55 of the Convention give a long list of the functions of ICAO. Principally they are divided into two groups: mandatory and permissive functions. The former are essentially obligatory, therefore, and the latter, those functions which they may undertake by choice. The former category includes its duties towards the Assembly (including submitting an annual report); duties concerning its own organisation and duties establishing other delegating bodies, in particular the Air Transport Committee and the Air Navigation Commission. In particular, the Council must request, collect, examine and publish information relating to the advancement of air navigation and the operation of international air services, including information about the costs of operation and particulars of subsidies paid to airlines from public funds (Article 54).

There is provision for the reporting of infractions of the Convention, first to contracting states and then the Assembly, where action has not been taken by the offending state within a reasonable time after notice of the infraction (being breaches or violations).

Also worthy of note, the Council is obliged to adopt (in accordance with Chapter VI of the Convention) international standards and recommended practices which, for convenience, are to be designated as Annexes to the Convention and notified to all contracting states.

This is, in effect, therefore the executive arm of ICAO dealing with the evolving practice of regulatory creation. The Council is also obliged to consider any matter relating to the Convention to which it is referred by any contracting state.

The permissive functions relate to a number of functions which the Council may carry out including delegation of work to the Air Navigation Commission and generally consider all important aspects of air transport.

The Air Navigation Commission

The Air Navigation Committee is an important sub-body working under the Council. Its establishment, constitution and duties are laid down by the original Convention. Its main function is to consider and recommend the adoption and modification of Annexes to the Convention. It can establish technical sub-commissions and advise the Council on the collection and communication of information to contracting states (which it considers necessary and useful for the advancement of air navigation) (Article 57).

Finances

It is the duty of the Council to submit annually to the Assembly budgets, statements of account and estimates of all receipts and expenditure. The Assembly votes the budgets with any modifications as appropriate. These expenses are then, normally (there are a few exceptions), apportioned among the contracting states on the basis that the Assembly shall from time to time determine (Article 61). There are penalties for non payment; the Assembly may suspend the voting power in the Assembly and in the Council of any contracting state that fails to discharge within a reasonable period its financial obligations to the organisation. Each contracting state must pay its own expenses including those of its own delegation, representatives and appointees.

In conclusion, therefore, ICAO is essentially self-financing by its members, the contracting states.

International arrangements

There are provisions for ICAO to make arrangements with other bodies and organisations in trying to achieve certain goals (Articles 64–66).

The provisions referred to are those within the Chicago Convention of 1944 or as since amended. It is important to be aware of the extensive work undertaken by ICAO since that time embodied in a sequence of Annexes (some 18 by 1991). These are related to a wide range of air transport issues covering such areas as the regulation of air navigation, aircraft, air traffic services, safety and accident investigation and various aeronautical matters. In particular we will see (in Part IV hereto) that Annexe 16 produced essential standards and regulations for the restriction of aircraft noise and engine emissions.

European Civil Aviation Conference

Having considered the role of ICAO, mention should be made of the European Civil Aviation Conference, known as ECAC. ECAC was essentially created by ICAO in 1954 at an ICAO conference in Strasbourg. The objectives for which ECAC was formed are fundamentally to review generally the development of European air transport in order to promote the co-ordination, the better utilisation and the orderly development of such air transport, and to consider any special problem that may arise in this field.

Its members are those states invited to be members at the 1954 Strasbourg Conference together with such other European states as ECAC may subsequently agree to admit. ECAC works closely with ICAO and performs a useful function in researching, reviewing and proposing issues concerning civil aviation. Recently, for instance, it has made a contribution to the debate on noise restrictions by producing suggested proposals for further restrictions to achieve Chapter II, and beyond, compliance (see Part IV on noise regulation generally).

It has also been involved in producing useful agreements and practice proposals. Of particular note, for instance, is the multilateral agreement on commercial rights of non-scheduled air services in Europe reached at Paris in 1956. This provides a set of rules between signatory states for the air traffic rights between those nations of certain categories of services.

International Trade Associations

International Air Transport Association

The International Air Transport Association (or IATA, as it is commonly known,) is essentially a trade association for international airlines. It is, therefore, a voluntary organisation established for the benefit of its members and is not in any sense a Government body. IATA is indeed based on the principle that the autonomy of its individual members is respected.

Though established in the 1940s, IATA evolved from a similar organisation. That earlier organisation was founded by a handful of airlines as early as 1919 when, even then, the benefits of established uniform practices and standards for the future development of civil aviation were appreciated.

The IATA of today was incorporated by a special act of the Canadian Parliament in 1945. Its Articles of Association can only be changed with the consent of the Canadian Government and it maintains a head office in Montreal. The objects of IATA, apart from being concerned generally with the interests of its members (as with any other such association), are based on a central goal of trying to achieve a unity in the operation of international airlines. Within this, its aim is to promote safe, regular and economical air transport, to foster air commerce and provide a means for collaboration among air transport enterprises and to co-operate with ICAO.

In trying to achieve these aims during the last, nearly, half a century, IATA involvement, influence and significance in the civil aviation industry should not be under-estimated. It has, through its regular general meetings, specialist committees and research projects, investigated and reported on many vital issues affecting both the operators and users of the industry. In many areas, such as standard form documentation, fares and interline ticketing and financing, it has established uniform systems and practices followed by most international airlines.

Membership is essentially for international airlines, originally limited to those operating scheduled services. Since 1974, however, when the rules were changed, membership has been open to certain categories of non-scheduled operators. There are two basic types of membership, 'active members' and 'associate members'.

Active members are essentially the international carriers, who have full voting rights at all meetings. Members in this category have, in effect, full membership.

Associate membership is essentially for domestic carriers (with certain qualifications) whose membership rights are restricted. For example, they do not have a right to vote at IATA meetings or at the traffic conferences.

Applications for membership should be made to the Executive Committee for approval. If that approval is rejected there is a right of appeal to the

General Meeting of IATA. Members can resign their membership at any time, which is effective 30 days after such notification has been given to the Director General. Membership can also be terminated involuntarily by IATA forcing a resignation. The Executive Committee can, for example, terminate membership for breach of the Articles of Association or a regulation or failure to comply with certain procedures. It should also be noted that membership is only open to airlines who (inter alia) are registered in a state which itself is a member of ICAO (the International Civil Aviation Organisation), see page 3 above. Hence, if that state ceases to be a member of ICAO then its airline(s) may in theory have to resign from IATA.

Constitution

Overall control of IATA is invested in the Annual General Meeting and this has three main roles. First, it controls the internal affairs of the organisation, such as budgeting, election of officers and the membership of various committees. Secondly, it deals with major problems and issues affecting the whole industry. In this respect it discusses and passes resolutions suggesting action to be taken by IATA, by various governments, and by other organisations. Thirdly, the general meetings are used informally for the airlines to meet each other and to discuss problems and matters of mutual interest and concern on a regular basis.

The Annual General Meeting can and does effectively control the structure of the Association. The resolutions passed by an Annual General Meeting are normally binding on all members except where, for example, such resolution contravenes that applicable law or regulation or policy of the state of which the member is a national.

The Annual General Meeting elects the President of the Association as well as members of the Executive Committee and the various standing committees, sub-committees and working parties.

The Executive Committee consists of 21 members who, in conjunction with the Director General and his Secretariat, suggest recommendations to the traffic conferences and to the general meeting. The Executive Committee also controls violations and tries to find ways of enforcing binding agreements. The standing committees meet once or twice a year but no action or decisions are binding upon its members until approved by the Executive Committee or the General Meeting.

The main committees are:

The Technical Committee – this is concerned with the safety and efficiency of flights.
The Financial Committee – this considers the standardisation of accounts and accounting procedures. It operates the important 'clearing house' system. Its administration is based in London and provides the means of settling on a monthly basis the debts and credits that arise between airlines

for services provided to each other. A typical example is interlining of tickets, where a ticket is bought by a passenger from one airline and is then exchanged for a ticket from another airline.

The Legal Committee – this is concerned with legal matters and the development of the law as it affects aviation.

The Medical Committee – this is concerned with the fitness of pilots and undertakes various studies on such medical subjects as the impact of distress on passengers.

The Traffic Advisory Committee – this advises on the traffic conference which deals with one of the most important areas of IATA's work. This considers such matters as cost, fares, rates, schedules, and agency matters and standardisation. Traffic conferences are held regularly and each conference deals with a particular geographical area. For this purpose the world is split into three areas being in general terms: 1) North and South America; 2) Europe (including parts of the Middle East) and Africa; and 3) Asia and Australia.

Each active member of IATA (i.e. International Airlines) is a voting member of the Traffic Conferences, but only participates as such in the areas in which it flies.

Two types of membership can attend traffic conferences: namely, the voting members, which are the relevant active members, and the non-voting members, which are the associate members or the active members who do not actually fly within the area of the conference (and are, therefore, not allowed to vote).

For example, an active member who flies services within Europe and between London and New York, say, would be a voting member of both the American and European conferences, but not of the Asian and Australian ones, which it could only attend as a non-voting member. Finally, of course, all members of traffic conferences must be members of IATA.

The person representing the voting member at the conference must be a primary traffic official. As a member organisation it is strictly governed by the Articles of Association of IATA. For example, that person must have full power and authority to bind a member in any matter.

Traffic conferences consider and act on all international traffic matters of concern to members. They are concerned, inter alia, with the analysis of operating costs, fares, rates, charges for passengers and cargo and so on.

The traffic conference agreements require the approval of each member's own Government. The approval, disapproval and/or reservation of each government is expressed in the administrative order. All resolutions are therefore classified into four types, in respect of the effectiveness of the resolutions after each government has approved or otherwise each resolution. The resolutions are classified by 'lettered' types – type A, B and so on. These classifications indicate the different status of the resolution according to each government: for example, that the resolution or part is void, or a conditional approval and so on. This can, of course, result in a complex and confusing picture. The effectiveness of a resolution, therefore, varies from country to country and it is difficult to achieve any overall uniformity.

Enforcement

There are obvious problems of enforcement with what is essentially a voluntary association of members and not a government body with the powers of legislative enforcement within a given jurisdiction. Clearly, though, where government approval is obtained this is normally accompanied by some sort of willingness to enforce those provisions where that government has jurisdiction. Over the years, though, government response and enforcement by the courts has been a difficult and controversial subject.

Within IATA, however, there are procedures for enforcement. After resolutions have been filed with, and approved by, the relevant governments, the members are obliged to observe those resolutions approved by their own government. A breach may result in a fine imposed by the IATA Compliance Commission. That enforcement may not, of course, be limited to IATA if approved by a government; for example, governments tend to enforce IATA resolutions when they form bilateral agreements in accordance with their own national law. Furthermore, IATA resolutions can be conceived as incorporated within the national laws of member states if that resolution is approved by the government concerned.

The Secretariat of IATA has its own enforcement office which investigates and punishes violators. The office acts upon the complaints of other members or at the Director General's discretion. If members are investigated they must furnish the investigator with information, records, and interviews. If this is refused, it is itself a breach. The investigator reports finally to the Director-General and if the member appears to have committed a breach, the Director-General will ask the member for a written reply to the allegations. On receipt of the reply the Director-General can then discuss the matter with the member concerned and/or refer the matter to the Commission for a hearing and thereafter a decision.

The IATA Compliance Commission consists of one commissioner who hears the case. The Commissioner's decision is notified to all members of the conference and of IATA. The offending member can be reprimanded, fined or ultimately expelled from IATA. There is a right of appeal to the Executive Committee where new evidence is available. If the Executive Committee is satisfied that there is new evidence to consider they can refer the matter to the Commission whose decision is then final.

IATA is obviously more than just a trade association. Though a private international organisation, it has often managed to obtain government backing and, in particular, support for and approval of some of its resolutions. It plays an active and, indeed, important role in the organisation and regulation of commercial civil aviation.

Other air transport associations

There are, of course, other associations of airlines and related interests. Most of these are on a regional basis such as the African Airlines Association (AFRAA), the Orient Airlines Association (OAA) and the Air Transport Association of America (ATA) to name but a few of the growing number of such associations of varying size and significance. Of particular note, in Europe, is the Association of European Airlines (AEA).

Association of European Airlines

The AEA is based in Brussels with a membership of over 20 European airlines. It obviously exists primarily to look after the interests of its members and in so doing has produced co-ordinated viewpoints on a number of vital issues affecting airlines.

This has involved the organisation in useful research and reporting activities on a number of matters affecting airlines. For example, it has been involved in the noise debate and represented the airline view on the introduction of various restrictions including undertaking a survey into the potential cost of banning Chapter Two aircraft among its European member airlines (see Part IV on noise regulation generally). These sorts of surveys produce an interesting and valuable service to the industry.

Eurocontrol

Eurocontrol is a European body established for the safe organisation of air navigation in specified airways above Europe. Its inception was in 1960 at the Brussels Convention. This international convention produced a treaty signed by six original signatories: namely, France, Luxembourg, Belgium, United Kingdom, Netherlands and Germany. It came into force in March 1963 and has since been adopted by other European countries. The original convention has also been amended on several occasions culminating in the Brussels Protocol of 1981.

In trying to achieve a common policy for the standardisation of regulations in matters of air navigation, the Convention was pursuing the aims of the International Civil Aviation Organisation as originally expressed by the Chicago Convention of 1944.

The Brussels Convention is then another example of international co-operation initiating regulation at an international level.

This is the first level of the familiar two-tier pattern arising from other conventions, such as Chicago in 1944 and Warsaw in 1929. The second stage is to give effect to those provisions by their adoption nationally by each state. This was done in the United Kingdom by the Civil Aviation (Eurocontrol)

Act of 1962. Though this Act has subsequently been repealed, the essential provisions relating to Eurocontrol are now found in the consolidating Civil Aviation Act of 1982 ('Act') and as subsequently amended by the Civil Aviation (Eurocontrol) Act of 1983.

The Convention is subtitled a 'Convention Relating to Co-operation for the Safety of Air Navigation'. For this purpose the contracting parties agree to strengthen their co-operation in matters of air navigation and, in particular, to provide for the common organisation of air traffic services in the upper air space (see Article 1 (1) Brussels Convention, 1960).

To achieve this they established under the 1960 Convention a 'European Organisation for the safety of air navigation' (Eurocontrol). This organisation was to consist of two parts: namely, a permanent commission for the safety of air navigation and, secondly, an air traffic services agency. For these purposes the expression 'air traffic' comprises civil aircraft and those military, customs and police aircraft which conform to the procedures of ICAO. The organisation is expressed to have legal personality: in other words, it is an independent body having a legal capacity in its own right. Eurocontrol is based in Brussels with agents in member countries. The following is an indication of some of the main provisions of the original Convention of 1960.

The Commission

The Commission is composed of two representatives of each contracting party. Each contracting party shall, however, have one vote. The general aims of the Commission are to promote, in co-operation with the national military authorities, the adoption of measures and the installation and operational facilities designed to ensure both the safety of air navigation and the orderly and rapid flow of air traffic within the defined air space of the contracting parties. For these purposes the Commission has various responsibilities. In general these include undertaking the necessary studies, promoting common policies in specified areas and, in particular, determining the configuration of the air space in respect of which the air traffic services are entrusted to the Agency. The Commission has various functions of a policy, financial and supervisory nature in respect of the Agency.

Constitution

The legislative process is essentially a three-stage procedure. First, *recommendations* are formulated by a majority of the members of the Commission. These are referred by their representatives to the individual member states for implementation. Secondly, for *decisions* the unanimous vote of the commission is required and once achieved they have a binding effect on each contracting party. Thirdly, there are *directives* of the Commission, which require a majority of the votes of the contracting parties (subject to complex weighting provisions).

The Commission shall establish its own rules of procedure which must be adopted unanimously (Article 10). It should be noted that the Commission (subject to the Convention) shall alone be empowered to conclude on behalf of the organisation those agreements with international organisations, member states of the organisation, or other states which are necessary for the execution of the tasks entrusted to it by the Convention and for the functioning of the organs established thereby or created for the purpose of its application (Article 12).

Agreements may be made between the Organisation and any state which is not a party to the Convention but wishes to use the services of the Agency (Article 13).

The Agency

The Agency is in effect the executive arm of the Organisation, and whereas the Commission carries out tasks of a constitutional nature, the Agency undertakes the day-to-day operation of tasks of providing air traffic services.

The Convention expressly states that the contracting parties shall entrust to the Agency the air traffic services in the defined air space (Article 14). The Agency controls the air space by giving all necessary instructions to aircraft commanders who must (excepting cases of force majeure as provided) comply with those instructions (Article 18). Infringements are recorded in reports by authorised officers (Article 19). The Agency shall provide tariffs and conditions of application of these charges for the users of the services it provides (subject to the approval of the Commission) (Article 20). These charges are then collected by Eurocontrol or its agents in member states. For example, the Civil Aviation Authority is, in effect, Eurocontrol's collecting agent in the United Kingdom.

Miscellaneous

The Organisation is, in general, exonerated from taxes, duties and various charges that would arise (as specified) in the terms of contracting states (Article 21). Executive provisions are set out in the Convention detailing the powers and duties of the Organisation in its respective organs, the Commission and the Agency. It should be noted that (under Article 39) the Convention was stated to remain in force, initially, for a period of 20 years. Subsequently the Brussels Protocol of 1981 made a number of amendments to the original Convention including prolonging its life by a further period of 20 years from the date of entry in force to the Protocol (with a provision for a further 5 years renewal). The above paragraphs summarise a few of the extensive provisions laid down by the original Brussels Convention of 1960.

Many changes were made to the drafting of the original Convention by the 1981 Protocol and the two texts should be read together.

Civil Aviation Act, 1982

The 1982 Act provides in Schedule 4 thereto for the recognition of Eurocontrol pursuant to the Brussels Convention (Section 24). The first provision in that

Schedule states that Eurocontrol shall have the legal capacity of a body corporate. This reflects and embodies the provision in the Convention which states that in the territory of the contracting parties the Organisation will have the fullest legal capacity for which corporate bodies are entitled under national law.

Eurocontrol is to be entitled to certain rights and privileges pursuant to the Schedule. For example, it is entitled to certain exemptions and reliefs in relation to taxes and similar charges. Section 73 of the Act provides that the Secretary of State may make regulations for requiring payment to him or the CAA or Eurocontrol of charges in respect of air navigation services which either in pursuance of international arrangements or otherwise are provided for aircraft by him, the CAA or Eurocontrol (or other persons).

Civil Aviation (Eurocontrol) Act, 1983

There have been further statutory provisions made more recently pursuant to the Civil Aviation (Eurocontrol) Act 1983. In particular, this made a number of amendments to the Civil Aviation Act of 1982. In general, these relate to two specific matters (the enforcement of foreign judgments in respect of route charges, see Section 1, the Civil Aviation (Eurocontrol) Act, 1983 and immunities and privileges of Eurocontrol, see Section 2 thereof) in respect of route charges and amenities and privileges of Eurocontrol.

Regulations, 1989

The main body of the Eurocontrol regulations, as they relate to the United Kingdom, are those laid down by statutory instrument made pursuant to the Act. The current regulations are found in the Civil Aviation (Route Charges for Navigation Services) regulations of 1989 (SI 1989, No. 303) as amended. These regulations came into force on 1st April 1989. All operators of aircraft should be familiar with these regulations.

In essence the operator of *any aircraft* (irrespective of the state in which it is registered) for which navigation services are made available (subject to the provisions of the regulations) in a specified air space shall pay the Organisation (being Eurocontrol) in respect of each such flight a charge for those services.

Liability, therefore, is placed in the first place on the person who is apparently the operator. The regulations provide, however, that where the Organisation is unable, after taking reasonable steps, to ascertain who is the operator, it may give notice to the owner of the aircraft that it will treat him as the operator until he establishes to the reasonable satisfaction of the Organisation that some other person is the operator. Until that happens, the Organisation is entitled to treat the owner as the operator (under the regulations). Apart from the power of detention and sale, this is another area where an unsuspecting owner or lessor may be vulnerable (see pp. 121 and 129 on the rights of detention and sale of aircraft).

These services do not, of course, include navigational services provided in connection with the use of an aerodrome, which are dealt with quite separately and covered by different regulations and charges (see Part II, p. 124).

The charge is calculated in accordance with the provisions set out in the regulations (in particular Regulations 5, 6 & 7). Those regulations explain in detail the calculation of the charge by reference to a formula taking into account various factors, including for example distance and weight factors, multiplied by a specified unit rate.

The amount charged is to be paid to the Organisation in ECUs, which are defined within the regulations as European Currency Units (Regulation 5 (1)) calculated in accordance with Regulation 8. The equivalent in sterling may be recovered in the court of competent jurisdiction in the United Kingdom (Regulation 5 (2)). Indeed, the Organisation has been known to take legal action in the United Kingdom courts for recovery of unpaid Eurocontrol charges. They have also been known to commence winding-up proceedings against a United Kingdom registered company for continued non-payment of Eurocontrol charges.

There have in practice been complaints, in particular by smaller operators, of the Eurocontrol system which, in the past, has appeared to some to be unnecessarily complicated and bureaucratic. Certainly some complained the charges submitted were not always correct and the necessary processing and validation of Eurocontrol charges was a burden on smaller operators. One reason for potential dispute has been whether or not the flight in question was within the category of exempt flights.

The regulations (Regulation 9) list a number of flights that are exempt. Some examples are flights by military aircraft or those for specific purposes such as search and rescue operations, instruction or testing of flight crew, checking or testing air navigation equipment, for certification of airworthiness qualifications and so on. Other exemptions involve geographical or weight restrictions. Significantly, there is an exemption for flights by aircraft of which the maximum total weight authorised is 5,700 Kg or less made entirely in accordance with the Visual Flight Rules of the Air and Air Traffic Control Regulations, 1985. There are other exemptions including certain flights in the United Kingdom by helicopter.

The Regulations also deal with the powers of detention and sale of aircraft for unpaid charges. These provisions and their relevance to operators and others with an interest in the aircraft is discussed in (Chapter 22 of Part II).

2 The European Economic Community

No mention of international institutions could be made without particular reference to the European Economic Community ('EEC'), Common Market or European Community as it is also known.

Since its inception in the 1950s, the EEC has grown into a formidable economic and political grouping of European nations. There are now (as at 1st January 1991) twelve member states: the original six; namely, France, Germany, Italy, Holland, Belgium and Luxembourg, joined by the United Kingdom, Denmark and Ireland in 1973, Greece in 1981, and Spain and Portugal in 1986.

As part of the increased economic co-operation working towards a unified Common Market by 1st January 1993, is the expanding area of Community legislation including the regulation of transport and, inevitably, involving aviation. A brief summary of the background to the Community and its institutions is given below.

Background

Some argue that a united Europe under some form of political and economic union is not only desirable, but a wholly-natural consequence of a common history, geography and civilisation. The crucial impetus has been the desire for peace.

The concept of the EEC was born in the aftermath of the second World War. For the second time this century, from 1939 to 1945, Europe was thrust into destructive and debilitating conflict. In the grim years that followed, while trying to rebuild the damage to flattened cities and revive industry and commerce, many yearned for some way of securing the future peace of Europe without fear of the wars that had been a feature of European history for centuries. It was also apparent that with modern technology, war was becoming increasingly horrific with far more devastating consequences for mankind.

In the late 1940s and early 1950s there was, therefore, an emergence of many hopes, aspirations and ideals which did produce new institutions aiming towards co-operation and peace between European nations. Most significantly there was the birth of what was to become the EEC. This did not emerge as a complete, single identity but evolved over the years. The original steps that laid the foundation for the European Community can be summarised as follows:

The Marshall Plan

Post 1945 it was clear that many countries in Europe needed financial assistance in order to rebuild their economies. Under the famous Marshall Plan, instigated by George Marshall, the United States Secretary of State, in 1947, America produced a package of aid to Europe. This led to the establishment of the Organisation for European Economic Co-operation (OEEC) in 1948, which later became the Organisation for Economic Co-operation and Development (OECD) in 1961. This laid the foundation of a centralised form of economic planning and co-operation.

Schuman Plan

In 1950 the French Foreign Minister, Robert Schuman, proposed a far-reaching plan of co-operation between his own country, France, and their time-old enemy, Germany. In the aftermath of the Second World War there were not only the usual concerns about future peace within Europe, but fears of the Communist force behind the Iron Curtain. Troubles in Czechoslovakia and Hungary followed by the 'Cold War' period alerted Europe to the need for Western defences and the threat from the East. Defence was an important issue yet Schuman had the vision to promote the defence of Europe by encouraging co-operation and, therefore, securing peace between its nations. The idea was to integrate coal, iron and steel industries. Steel was essential for war, but steel could not be made without coal. By harnassing the joint resources to a common goal, peace within Europe was more secure and the threat of foreign invasion less likely.

Thus, France and Germany, followed by Italy, Belgium, Holland and Luxembourg became members of the European Coal and Steel Community. This was established by treaty signed in Paris in April 1951.

A common market

In 1957 two momentous steps were taken towards further co-operation and unity. By two treaties signed in Rome in March of that year, two important community institutions were established. First was the European Atomic Energy Community, known as ERATOM. Secondly, the European Economic Community was established. This Treaty of Rome provided for a wider basis of economic co-operation with far-reaching goals, through which a European Community could develop in the ways we see today.

By the late 1950s, then, there were three distinct organisations each set up by treaty and each with its own organisation and regulations: namely,

(1) European Coal and Steel Community (ECSC),
(2) European Atomic Energy Community (ERATOM), and
(3) European Economic Community (EEC).

Merger Treaty

Further progress was made in 1965, when, by a treaty known as the 'Merger Treaty', agreed in April of that year, attempts were made to co-ordinate and unify the workings of the hiterto three very separate organisations of the Community. The Treaty, to be effective in July of 1967, provided for a single Council and Commission (where they had previously been duplicated) so that all the internal institutions of the three organisations could be consolidated into single entities. This was a major step forward in progressing the mechanics of the Community.

Towards 1992

More recently, during the 1980s, considerable progress has been made towards achieving a Single Common Market in the spirit of the Treaty of Rome of 1957. The means to achieve this was set out in a white paper published by the Commission in 1985. This was followed by the Single European Act signed by all the Heads of Government of Member States, in 1986, which committed them to a Single European Market by 1st January 1993. This has meant an energetic programme of action to achieve that goal.

Change in the last few years has therefore been rapid and significant including steps to remove physical, technical and legal barriers and enable the free movement of peoples, goods and capital in a competitive and commercial environment restricted only, where possible, by common standards and regulations. There has been, therefore, an influx of community legislation in the late 1980s and lively debates as to how far economic and even political union should be realised in the future. In any event, by 1993 there should be in Europe a Single Open Market of some 325 million people, forming the world's leading trading power.

EEC institutions

As a body the European Community consists of four main institutions. In effect they cover the legislative, executive, legal and democratic functions of the Community. A brief summary of each is given below.

The European Parliament

The Parliament is located in Strasbourg and is the democratic forum of the Community; its representatives being, since 1979, directly elected by the citizens of Member States. Candidates are elected by proportional representation in all states except the United Kingdom, where the single-ballot majority-vote by constituency system is used. The first of these direct elections was in June 1979, and the second in June of 1984. There are now (as at 1st January 1991) twelve member states electing a total of 518 MEPs (Members of the European Parliament) with each of Germany, France, Italy and the United Kingdom having 81 MEPs; Spain, 60; Holland, 25; Belgium, Greece and Portugal, 24; Denmark, 16; Ireland, 15; and finally, Luxembourg, 6.

Parliament sits normally for one week each month (with a break in August), and the MEPs are grouped not by nationality but by political association, ranging from the far left to the right-wing parties. The function of the Parliament is to directly represent the people, as opposed to the Council representing Governments, and both are independent bodies as opposed to the Commission. Indeed, the independence of the Parliament has sometimes resulted in disagreements between it and, say, the Council.

Despite the Parliament being the directly-elected body, it is not, like most national parliaments, the main legislative body, which as we see below is the role of the Council of Ministers. The main role of the Parliament under the Treaty of Rome is as a consultative body enabling it to debate and voice an opinion on many issues. It is not, however, primarily a decision-making body as is the Council of Ministers; where the Parliament does exert considerable influence is in the area of budgetary control. Finally, though the Parliament is independent of the Commission, the Commission is responsible to it and can be dismissed by it.

The Council of Ministers

The Council of Ministers is situated in Brussels, where it meets several times a year. As the name suggests, the Council is composed of Government ministers representing each Member State. Normally, the Foreign Minister of each State is the representative, though this depends on the nature of the matter in hand. For instance, on agricultural issues the Minister of Agriculture from each respective government would attend. The Council appoints its own President from one of its members who holds the position for 6 months, thus allowing for frequent rotation of the post.

The voting mechanism, whilst taking account of the voice of each Member State, tries to balance the difference in their relative sizes. Thus, the number of votes given in respect of each state varies accordingly: from the smallest member, Luxembourg, which has two votes, to the larger countries, which have ten votes each. The voting arrangements have changed over the years with the increased membership and so also has the emphasis on majority voting for decisions. Many more decisions can now be taken by majority vote (in order to speed up the effectiveness of the Council), with matters requiring a unanimous vote restricted to special situations (for example, for certain matters of vital national interest).

The Members of the Council being Ministers from each Member State are, as Government representatives, responsible to their respective Governments and not to the European Parliament (unlike the Commission, see below). The Council is, therefore, very much a political forum representing particular political parties (being the elected Governments) of Member States. In that sense, the members of the Council are independent voices speaking for Governments and not interests within the Community. This is in contrast to the Commission whose members have a duty to act first and foremost in the interests of the EEC and not as a representative of any particular Government or State.

The Council is, therefore, a vital link between Governments and their interests and the Community. Sometimes, because of the different interests represented, the Council does not see eye to eye with the Commission. In order not to let the more partisan interests of the Council become too paramount there have been more recently some attempts to limit the intervention by the Council in the general running of the Community.

Traditionally, however, the Council has been the major decision-making body within the Community. Its essential function is that of the legislature (rather than Parliament as might otherwise have been supposed) whereas in general terms the Commission is the executive arm and the European Court, the judiciary.

The European Commission

The Commission, with its headquarters in Brussels, is in effect the executive arm of the Community. It is a large, permanent body dealing with the extensive and extending day-to-day administration. It is in many senses the heart of the Community undertaking many vital functions not least of which is to research and then make proposals for the further development of Community legislation which it then lays before the Council.

The members of the Commission, or Commissioners, are appointed by agreement between member governments for a four-year term of office. They are at present (as at 1st January 1991) 17 in number, each one obliged under oath to act independently in the best interests of the Community as a whole. They must be, therefore, independent of national Governments and of the Council. The Commissioners and their staff are in effect the full-time career civil servants of the Community. As such the day-to-day control and general influence of the Commission is considerable. Though in theory Parliament has the power to make the Commission resign, the Commission has considerable strength and substantial influence. It has some direct legislative powers (which are not solely restricted to the Council), certain supervisory or regulatory roles, as well as administrative functions (being in effect the civil service of the Community), and an important role in proposing legislation for consideration by the Council of Ministers. In this role it initiates and influences Community policy and interests. Under the Commission a number of committees are instigated to manage, advise and regulate.

The European Court of Justice

When looking at the basic concepts of law discussed in Part III of this book, on Legal Liability, (see p. 139) it will be seen that any system of law within a given jurisdiction is of little effect unless there is a judiciary empowered to enforce that body of law. Thus, within the European Community there has to be a way of enforcing its laws amongst Members States.

An essential institution, therefore, is the European Court of Justice which sits in Luxembourg and provides for the enforcement of the Community Law. The Court consists of 13 judges, each appointed for a 6 year term of office by agreement amongst member states. The judges are also assisted by Advocates-General.

The Court has considerable authority in enforcing Community Law which includes the Treaties (particularly the clarification, implementation, and interpretation of the Treaty of Rome) and the increasing secondary legislation made pursuant thereto. The Court is widely accessible and many actions have been brought, for example by individuals, Governments, and the Commission. It is bound to play an increasingly significant role in the life of the Community.

Finally, it should be remembered that, despite the legal debates that have surrounded what is a complex subject affecting sensitive issues of national sovereignty, in general terms Community Law is supreme and takes precedence over the law of Member States.

Aviation and the European Community

Inevitably, aviation policy and regulation has been a part of the development of the European Community.

In the early years, however, aviation issues were given little consideration. Indeed, there was some question as to whether air transport was covered by, and therefore subject to, the Treaty of Rome. The confusion arose primarily because of a reference to transport which specified road, rail, and inland waterways without specific reference to air transport. However, this debate is now long over and it has been gradually accepted (with the help of several clarifying judgments by the European Court) that in general the Treaty of Rome does apply to aviation.

This opened up a massive area of potential debate about the possibilities and desirabilities of centralised Community regulation of air transport. The extent of Community intervention is not an easy question to justify or assess. It touches on many complicated issues, such as the national sovereignty of air space, the numerous international traffic-right agreements between individual Governments and non-member states, a state's involvement in its own carriers and national obligations to a variety of international treaties affecting aviation. Perhaps the crux of these and similar issues is an individual state's established sovereignty which affects so many different areas of air transport. The question is, then, how and to what extent can the Community superimpose itself on this international web of relationships and established national rights?

In the past, intervention by the Community has tended to be in the context of other general goals. For example, much of the activity in aviation is set against the principles of competition laid down by the Treaty of Rome. Thus, during the last decade we have seen intervention in such areas as fares and accessibility to routes and airports. Clearly, the Commission is anxious to attack the many monopolistic practices that have grown up over the years as a natural feature of air transport such as price-fixing arrangements, pooling agreements, and many restricted practices adopted under the guise of bilateral traffic-right agreements.

In this context it will require fairly radical reform to create within the Community a single, open, competitive market place for air transport services. Obviously the philosophy behind this accords with the main aims and ideals of the Treaty of Rome to create a single market with non-discriminatory access and free movement of money, goods and people within the Community.

Another general area of Community concern is the environment, and here too it has intervened in an aviation context to apply rules for improving noise standards of aircraft. This is referred to in more detail in the discussion on noise regulation in Chapter 37, Part IV of this book. Other areas, such as product liability (see Chapter 27, Part III) and denied-boarding regulations (see Chapter 30 in Part III), are similary discussed at appropriate points in the book. There has been regulation on a number of other areas such as competition (in relation to fares, inter-airline arrangements, airport accessibility etc.), traffic rights and technical matters which are not dealt with in detail in this book as they encroach on far larger topics.

The aviation activities of the Community can be traced through a series of stages from the late 1970s. This started first with a desire by the Commission to introduce a Community approach to aviation as part of a unified transport policy. To achieve the goals of the Treaty of Rome by creating a Single European Market with free movement of goods, a united and centralised transport policy is essential.

To this end, the Commission has produced a number of memoranda expressing its views and proposals. The first of these was in 1979 (Contribution of the European Communities to the Development of Air Transport Services – Memorandum of the Commission, E.C. Bulletin Supp. 5/79) which was in many ways a momentous event in the history of European aviation regulation and development. It in fact led to a relatively small amount of legislation but its impact in opening up the whole European debate was considerable. It heralded a new era of potential deregulation and liberalisation in European aviation. The timing of the 1979 memorandum was, of course, just after the considerable reforms in the United States of America in which their civil air transport system was deregulated and many of the regulatory functions of the Civil Aeronautics Board removed. Restrictions on entrance into the market place, routes and fares were kept to a minimum, normally to situations dictated by safety and security considerations.

Whilst the EEC Commission was undoubtedly aware of and even influenced by this new tide of deregulation, its methods were never meant to be a mirror image of the American experience. Liberalisation in European air transport has developed a different emphasis. It is based not so much on simple deregulation, that is the removal of all regulatory regulations and restrictions to a minimum, as to secure (by regulatory controls where necessary) a liberal environment based on free trade, fair competition and equal opportunity.

In 1984 the Commission produced a second memorandum (Memorandum No. 2 – Progress towards the Development of a Community Air Transport Policy in COM (84) 72 – Final) which stated its purpose as to develop and expand on the objectives of its earlier memorandum of 1979. It recognised the complex relationships in existence between individual Member States and third parties, (i.e. those states outside the EEC) and therefore concentrated on effecting change in the operation of air transport between EEC states. This was, of course, to be within the wider context of working towards an internal market in the Community.

Interestingly, whilst paying homage – within its own ideals – to liberalisation and therefore to some extent deregulation, the point is made that 'American-style deregulation would not work in the present European context'. Thus, the American experience is not to be a blue-print for Europe.

The Memorandum examines the reactions to the Commission's first Memorandum, developments since that time and produces proposals for future changes. An important reason for the first Memorandum was to initiate the debate on a Community air transport policy. This it had certainly done and though only a limited amount of actual legislation had resulted it instigated much discussion, research and proposals for the future.

The 1984 Memorandum provoked more significant changes in what has become known as the 'first package'. This included the Council Directive (Reference: 87/601) and a Council Decision (Reference: 87/602), both agreed by the Council on the 14th December 1987, to be in effect from 1st January 1988. In very general terms, the first related to fares for scheduled air services between Member States and the second to the sharing of passenger capacity between air carriers on scheduled air services between Member States and on access for air carriers to scheduled air-service routes between Member States.

Both these, the Directive and the Decision, have since been revoked and in effect replaced by measures in a 'second package'.

The 'second package' was largely a result of the Commission's proposals to amend the 'first package' and included the following two Regulations (which replaced the Council Directive 87/601 and Council Decision 87/602, respectively, referred to above):

(1) Council Regulation (EEC) No. 2342/90 of the 24th July 1990 on fares for scheduled air services, and

(2) Council Regulation (EEC) No. 2343/90 of 24th July 1990 on access for air carriers to scheduled intra-Community air service routes and the sharing of passenger capacity between air carriers on scheduled air services between Member States.

Both are effective from 1st November 1990 and both are stated to be due for change by 30th June 1992. The Commission was obliged to submit its proposals for that review by 31st May 1991. Clearly the amendments due to be effected by June 1992 are intended to be the final stage in a developing process towards creating the single market goal of 1st January 1993.

The effect of these present regulations of July 1990 is very broadly to regulate within certain parameters the fixing of air fares by establishing a filing procedure and system of approval or disapproval between Member States. The aim is to allow for reasonable freedom for carriers to set their fares subject to certain objective criteria. Whilst the relevant states can disapprove, their right to do so is restricted. For example, they cannot necessarily disapprove simply because the proposed fare is lower than others on the same route: thus, protecting the essential principle of competition whilst maintaining some checks on the extremes of a totally unregulated market, such as predatory pricing. The regulations are reasonably detailed and complex but overall they introduce some flexibility in what was previously a rigid structure.

In very broad terms, Council Regulation 2343/90 is concerned with first, access to the market for Community air carriers and secondly, the sharing of passenger capacity between air carriers in one Member State and the air carriers licensed in another Member State on scheduled air services between those states. There are a number of special cases which are excluded, but in general a system of regulation is established to allow within its parameters certain categories of air services by Community air carriers. There are detailed provisions relating to the different services, for example, third and fourth freedom flights between designated EEC airports to which access is to be on a reasonably unrestricted basis. Underlining the measures within this regulation is of course the framework of the Treaty of Rome and the creation of a Community market for the future.

Of particular note in recent years are the following:-

(1) Council Regulation (EEC) No. 2344/90 of 24th July 1990 amending Regulation (EEC) No. 3976/87 on the application of Article 85 (3) of the Treaty

to certain categories of agreements and concerted practices in the air transport sector;

(2) Commission Regulation (EEC) No. 82/91 of 5th December 1990 on the application of Article 85 (3) of the Treaty to certain categories of agreements, decisions and concerted practices concerning ground handling services;

(3) Commission Regulation (EEC) No. 83/91 of 5th December 1990 on the application of Article 85 (3) of the Treaty to certain categories of agreements between undertakings relating to computer reservation systems for air transport services;

(4) Commission Regulation (EEC) No. 84/91 of 5th December 1990 on the application of Article 85 (3) of the Treaty to certain categories of agreements, decisions and concerted practices concerning joint planning and co-ordination of capacity, consultations on passenger and cargo tariff rates on scheduled air services and slot allocation at airports;

(5) Council Regulation (EEC) No. 294/91 of 4th February 1991 on the operation of air cargo services between Member States;

(6) Council Regulation (EEC) No. 295/91 of 4th February 1991 establishing common rules for denied-boarding compensation system in scheduled air transport; (see p. 186, Chapter 30, Part III).

As to the future it is clear we live in times of rapid change – one need only look back through the last 20 years from the early 1970s – it was not until the 1980s that any significant intervention was achieved and the increase in the last 5 years has been dramatic. 1992 will undoubtedly see many more changes in preparation for the single European market of 1993.

It has been far from clear exactly how and to what extent the Community institutions would exert direct control over European air transport and thereby assume authority over areas hitherto under national control. In many instances, particularly such as traffic rights, this directly affects the national sovereignty of Member States. By tradition Governments have exerted an unusual amount of control in the area of air transport, as opposed to most commercial activities. Many, to a varying extent, own and control their 'flag carriers'. They have handed out licences for routes at their discretion; controlled capacity, fares and enforced monopolistic practices. How far these national rights are to be eroded and replaced by a Community authority is of great concern to many.

Certainly, this should become a lot clearer during 1992. Meanwhile, we can watch carefully a number of areas of potential or further change. We have already mentioned the Government-sensitive area of international traffic rights, incorporating a Government's fundamental right in exercising its nation's power of sovereignty to negotiate with other Governments what air services they will allow the other to operate within their respective sovereign air space. The question has already arisen as to whether all such Government rights should be transferred and divested in a central EEC authority which controls traffic rights within the Community and more significantly those between the Community and other non-Member Nations. So far the retention of some form of individual control by a Member State over its bilateral arrangements with non-Member States, at least, has been accepted, but whether this will continue remains to be seen.

Other areas of national control are also under discussion. For example, individual governments guard jealously the control they exert via economic restrictions exercised through financial and reporting requirements and, of course, route licensing. Will there be a Community Licensing Authority? Some idealists in Brussels would no doubt see this as the ultimate necessity for a Single European Market. They are likely, however, to meet with entrenched national opposition.

These sorts of challenges jeopardise the role, function and even existence of national control normally exercised through civil aviation bodies, notably the Civil Aviation Authority in the United Kingdom. It also brings to question the many other areas over which these bodies have regulatory control. One key area is that pertaining to safety, which affects a wide spectrum of regulation from crews to aircraft airworthiness. In both these areas discussions have taken place and proposals been made for centralising Community standards, requirements and regulations. Certainly on a practical level there is much to commend this as it would facilitate the proper freedom of movement of persons and trade envisaged by the ideals of the Single Market. Other areas that could be centralised and removed from national control are aircraft registration and ancillary regulations relating to aircraft mortgages, documentation and so on.

In the general commercial field much has been done, and a great more will be done, to create a more liberal market with easier access to new entrants, more competition and a gradual breakdown of monopolistic practices, though this may develop over some years. An interesting point concerning new entrants to the market raises the question of nationality qualifications, at present a normal requirement in most states for the licensing of their airlines. It is now generally accepted that some form of Community qualification will need to replace the strictly national criteria existing at present. This will mean that airlines of Community States will probably have to be owned and controlled to some extent not by nationals of particular states but by Community nationals. The Community will also take a special interest in the restructuring and merging of existing airlines. It has already shown its intentions and existing powers in this regard, for example in its investigation of the British Airways/British Caledonian merger in the United Kingdom.

Other areas of likely change are those affecting boundaries involving customs, immigration, security and other issues. This also brings in the matter of VAT and the possibility of unifying taxes. There have also been proposals to do away with duty-free goods when the Customs barriers are removed between European nations. This has brought about considerable opposition from those interests who at present benefit from the sale of duty-free goods at airports.

Airports are a major subject for Community intervention. Many of their traditional functions involving, for example, allotting slots, control of capacity, ground handling and other services have been embedded in restrictive-type practices. Already the Community has made some headway in reforming these areas, but it has a very long way to go before there is a truly open market in these services. One of the major problems is that of slot allocation where the issues are too complex and the actual slot availability too restrictive, (both in terms of available air space, traffic control and airport capacity) to allow a truly free market in such services. At the end of the day some criteria have to be in force to allot these vital and valuable rights.

One can also look at many other areas such as consumer protection and the environment. The recent denied-boarding regulations are an example of the former. In terms of the environment there will inevitably be further changes affecting, in particular, aircraft noise and emissions.

The very latest proposals under discussion reveal a fairly ambitious package aimed at achieving an internal market by 1st January 1993. In general, community carriers would be granted 3rd, 4th and 5th freedom (and potentially cabotage) rights within the community.

It is proposed that route licences and Air Operator's Certificates be granted on a national basis but in accordance with community requirements and procedures as regards fares, the process towards free pricing, should continue, to be fully implemented when normal competitive conditions have been established.

So, there is great scope for further change. How much will happen, and how quickly, will be revealed as we approach 1st January 1993. However, that will, no doubt, be not the end but the beginning of a new era, a new Europe and a new world of air transport.

3 National Organisation of Civil Aviation

The United Kingdom

At government level, the Secretary of State for Transport is the government minister responsible for civil aviation. The Secretary presides over the Department of Transport which is directly involved in a number of specific but limited areas of civil aviation. As we will see, in the UK, the regulation and supervision of civil aviation is largely the function of the Civil Aviation Authority (CAA). The role of the CAA is reviewed in the following chapter, p. 33.

The main source of aviation regulation in the United Kingdom is statute law, created by Acts of Parliament. Most of this statute law is to be found in the Civil Aviation Act 1982 which, in particular, lays down the roles of the main authorities for the control and regulation of civil aviation in the UK, being in the main covered by the Secretary of State (for Transport), the Department of Transport and, of course, the CAA. The Act deals with general issues of roles, functions, duties and policies. Pursuant to its authority numerous further detailed regulations are made, normally under statutory instrument.

Statutory law is not exclusively provided by the Act. Other legislation affects the numerous aspects of civil aviation regulation, for example, those enacting the provisions of international conventions such as the Warsaw Convention dealing with liability for international carriage enforced by the Carriage by Air Act 1961. Another example dealing with a quite separate but vital function of civil aviation is the Airport Act 1986, the source of most of the regulation of various airports in the UK.

The Secretary of State for Transport

The Secretary of State for Transport is, in general terms, the individual ultimately responsible to the government of the day for many aspects of civil aviation, although many of the supervisory and even regulatory functions have been delegated, in particular to the CAA.

Section 1 of the Civil Aviation Act 1982 states that the Secretary of State shall continue to be charged with the general duty of organising, carrying out and encouraging measures for specified areas of civil aviation, namely:

(1) the development of civil aviation;
(2) the designing, development and production of civil aircraft;
(3) the promotion of safety and efficiency in the use thereof;
(4) research into questions relating to air navigation.

These are essentially general functions of a policy and primary nature and do not involve the day-to-day regulation of aircraft operations. The Act also makes it clear that the Secretary of State is not empowered to authorise the production of civil aircraft (Section 1 (2)). On the contrary, the acquisition and disposal of aircraft, aero-engines and aviation equipment in discharge of the Secretary of State's duty under item (1) above shall be subject to the approval of the Treasury, thus featuring another government department and the Chancellor of the Exchequer in the diverse arena of civil aviation regulation.

Further to these general functions, the Secretary of State has a number of specific responsibilities specified by the Act which cover different areas of regulation. Many of these will be dealt with later, at the appropriate place in the book.

The Department of Transport

The work of various government departments may at one time or another impinge in some way on civil aviation. The Department of Transport is, of course, the most obvious example. Though many of the daily functions of the regulation of civil aviation are the domain of the CAA, certain key areas are dealt with by the Department of Transport. First, as a government department, it deals with the negotiation of international traffic rights, often referred to as bilateral agreements. In this situation the government represents the sovereign state in negotiating and reaching agreement with another sovereign state. On these matters of international relations it is clearly appropriate that a government department is responsible for these functions. Second, the Department of Transport is directly involved in certain safety matters, particularly the investigation of aircraft accidents.

Third, it is the Department of Transport which plays an important policy-making role in the area of aircraft noise in which it has been very active in the 1980s and, increasingly now, in the area of aircraft emissions and other aspects of aviation pollution. Other areas of Department of Transport involvement include, for example, national airport development policy and related issues, and the general co-ordination of aircraft security.

Other government departments and agencies

The wide spectrum of civil aviation can impinge on many different areas of public life, and therefore government. Most government departments are at some time, in some way, involved in matters affecting aviation. Apart from the Department of Transport and Treasury other government departments involved are the Department of Environment, the Foreign Office and the Department of Health and Social Security. This is due to the fact that many aspects of public interest and concern can arise out of aircraft operations.

We should also remember that although in this book we consider in some detail the work of the CAA, and indeed, it is the CAA which is responsible for the vast majority of regulatory functions of civil aviation in the United Kingdom, there are other non-central governmental bodies that also take an active role, particularly at a local level.

Local organisations

At a local level the most obvious bodies are local governments and airport authorities. In some instances the local government may itself also be an airport authority. In any event the local authority in whose area an airport is sited will to a variable extent, depending on the size and function of that airport, be involved in its operation on a number of issues, for example matters of land use and public health. The airport authority (whether a local authority or other body) is subject to considerable regulation made pursuant to various legislation including the Civil Aviation Act 1982, but more particularly the Airport Act 1986. Most of the major airports in the UK are now owned by a publicly-quoted corporation, the British Airports Authority plc or its subsidiaries. The regulation of airports is a vital area of civil aviation, but one which itself could be the subject of a separate book and it is not, therefore, considered here in any detail.

4 The Civil Aviation Authority

No body has more influence on civil aviation in the UK than the Civil Aviation Authority (CAA), an independent body corporate set up by statute to supervise and carry out many of the key regulatory functions in civil aviation.

The CAA is a relatively recent organisation set up in 1971. Prior to that date the regulation of civil aviation was handled by a series of government ministries and departments. Between the World Wars, responsibility for civil aviation was largely dominated by the Air Ministry, after which its functions were mostly continued by the Ministry of Civil Aviation set up in 1945. During the 1950s and 1960s further changes were made with the transfer of various functions and responsibilities involving a number of ministries, and eventually the Board of Trade in the late 1960s. Thus, through its regulatory history since the early 1920s, much of civil aviation in this country has been directly controlled and regulated by a government department. All this was to change in the early 1970s.

The turning point was the Edwards Report, entitled *British Air Transport in the Seventies* (Cmnd 4018), which was commissioned by the Government and published in 1969. A committee was set up, chaired by Sir Ronald Edwards, to enquire into and report on Britain's civil air transport. It is a lengthy, comprehensive and far-reaching document in its research and conclusions. In many ways its analysis and proposals laid the foundation for the regulation, organisation and supervision of civil aviation in modern Britain. One of its recommendations was the establishment of an independent civil aviation authority. Consequently, the momentous Civil Aviation Act 1971 made provision for the establishment of the Civil Aviation Authority which was formally set up in December 1971. Originally under the 1971 Act, the Secretary of State could give guidance to the CAA, thus ultimately making the Secretary responsible for policy formation, and the CAA responsible for carrying out that policy. This has since changed (the guidance provision being repealed by the Civil Aviation Act 1980) and policy formation (within the requirements of the Civil Aviation Act 1982) is now the responsibility of the CAA.

The Civil Aviation Act 1982 is a consolidating Act thereby repealing earlier legislation and it forms the current legal framework for the regulation of civil aviation in this country. It restates the constitution and functions of the CAA, most of which is dealt with in Part I of the Act. The remaining parts deal with regulatory provisions for which the CAA is primarily responsible and these are mentioned further below.

The objectives and criteria upon which the CAA is to base its decisions and generally carry out its duties and functions are specified at various points within the Act when dealing with the matters to which they relate. There are, however, some general objectives stated in Part I of the Act that are worth stating now and should be borne in mind at all times by the CAA and those dealing with it.

It is the duty of the CAA to perform the functions conferred on it in the manner which it considers best calculated:

(a) to secure that British airlines provide air transport services which satisfy all substantial categories of public demand (so far as British airlines may reasonably be expected to provide such services) at the lowest charges consistent with a high standard of safety in operating the services and an economic return to efficient operators on the sums invested in providing the services and with securing the sound development of the civil air transport industry of the United Kingdom; and

(b) to further the reasonable interests of the users of air transport services.

(See Section 4 (1))

In this context, British airlines (see Section 4 (2)) means an undertaking having power to provide air transport services and appearing to the CAA to have its principal place of business in the UK, the Channel Islands or the Isle of Man, and to be controlled by persons who are either United Kingdom nationals or are, for the time being, approved by the Secretary of State for the purposes of this subsection.

These overall general objectives are the fundamental key to the philosophy behind the regulation of UK civil aviation and the functioning of the CAA. The Act deals substantially with the CAA from its constitution to its statutory functions and duties which form the basis of the regulatory framework of UK civil aviation. Any operator or person involved in civil aviation is best advised to review the Act. It is reasonably clear in its drafting and covers many essential matters. In brief, the Act is divided into five main parts followed by a number of Schedules elaborating on various matters dealt with in the main text of the Act.

- Part I deals with the CAA itself, from its constitution and its basic functions to its financial and other administrative matters.
- Part II deals with aerodromes and land matters (including the various powers and rights vested in the CAA to acquire and otherwise dispose of certain interests in land for specified purposes).
- Part III headed 'Regulation of Civil Aviation' deals with various important regulatory matters some of which are reviewed in more detail in Parts II, III and IV of this book. In particular, Part III of the Act deals with areas of economic regulation, most notably air transport licensing. It also deals

with the provision of air navigation services and trespass, noise, and other rights affected by aircraft.
- Part IV headed 'Aircraft' deals with various aspects of aircraft. In particular it includes rights in and against aircraft such as mortgages, detention, sale and seizure.
- Part V is a miscellaneous and general section dealing mainly with the application of the Act and matters of interpretation.

In constitutional terms, the 1982 Act states that the CAA is to continue as a body corporate (as set up by the 1971 Act) (Section 2 (1)). It is to have a minimum of six and a maximum of sixteen members who are to be appointed by the Secretary of State who shall also appoint one such member to be the Chairman. The Secretary of State may also appoint a further one, or at most two, members to be Deputy Chairman of the CAA (Section 2 (2)). For the avoidance of doubt, it is stated that the CAA is not to be regarded as a servant or agent of the Crown, nor does it enjoy any such special status, privilege or immunity of the Crown. Its property is not owned by the Crown and it is a constitutionally separate and independent body. Thus, contrary to the view of some people, the employees of the CAA are not public servants. Furthermore, the CAA does not benefit from special tax privileges or exemptions and its profits are assessable to taxation in the normal way (Section 2 (4)).

Commercially, the CAA is an independent organisation with a stated financial objective of recovering the whole of its costs and achieving a reasonable return on its capital (see booklet produced by the CAA, *Britain's Civil Aviation Authority*, CAA December 225). Essentially, its costs are to be covered by the charges it makes to the users of civil aviation. The Act states (see Section 8 (1)) that it is the duty of the CAA to conduct its affairs to secure that its revenue is 'not less than sufficient' to meet charges properly chargeable to the revenue account, taking one year with another. In simple terms, its revenue should cover costs in each financial year. There is a statutory obligation (see Section 15) to keep proper accounts and accounting records, and for each accounting year to prepare a statement of accounts which with any report of the auditors must be sent to the Secretary of State who shall in due course lay them before the Houses of Parliament (see Section 15).

The Act deals with various other aspects of the CAA's finances such as its funding and borrowing powers. In terms of its revenue, it is given a specific right (after consultation with the Secretary of State) to determine charges which are to be paid to it in respect of the performance of its various functions (see Section 11). In practice, examples are as diverse as licence and certification fees, air navigation service charges and so on.

Having dealt with the CAA's basic constitution, the Act then prescribes its basic functions. Apart from matters relating to the CAA as an organisation (dealt with in Part I of the Act), and further functions conferred for the time being by virtue of the Act or other enactments, its main functions (see Section

3) are those relating to the following:

(1) licensing of air transport;
(2) licensing of the provision of accommodation in aircaft;
(3) provision of air navigation services;
(4) operation of aerodromes;
(5) provision of assistance and information;
(6) registration of aircraft;
(7) safety of air navigation and aircraft (including airworthiness);
(8) control of air traffic;
(9) certification of operators of aircraft; and
(10) licensing of air crews and aerodromes.

Part III of the Act headed 'Regulation of Civil Aviation' expands on many of these functions. It begins by giving the power to give effect to the Chicago Convention and to regulate air navigation. Under Section 60, Orders in Council may make such provisions as authorised by the Act. An air navigation order may contain such provisions as appears (to Her Majesty in Council) to be requisite or expedient for, in particular, carrying out the Chicago Convention (and any annexe or amendment thereof) and generally to regulate air navigation (see Section 60 (1) and (2)).

Without prejudice to the generality of these provisions, the Act (see Section 60 (3)) then itemises provisions that an air navigation order may contain. These matters cover several pages of the Act and indicate the many and various areas of regulation. For example, in terms of aircraft we see reference to registration and airworthiness certification. These are vital areas covered by the current Air Navigation Order 1989. Airports are another essential area mentioned and some of the services found there such as Customs and Excise and air naviagation services. More unusual matters are also mentioned such as noise restrictions and provisions for regulating or prohibiting the flight of supersonic aircraft over the United Kingdom.

Policy statements

The CAA has a statutory duty (Section 69 (1)) to *publish* from time to time a statement of the policies it intends to adopt in performing certain of its statutory functions in relation to air transport licensing. Furthermore, if it considers it appropriate, the Secretary of State may by notice in writing require the CAA to publish a statement of the policy it intends to adopt (in performing particular functions) (Section 69 (2)). The CAA is obliged to publish the statement within six months of the date of the notice.

Before publishing any such statement, the CAA is obliged to *consult* certain persons. These are persons that appear to the CAA to be representative of two groups: namely, the civil air transport industry of the UK and users of air transport services. The manner of publication of a statement should be as the CAA may determine (Section 69 (4)).

Publication

Any notice or other matter required by the Civil Aviation Authority Regulations or, where appropriate, the Act, is to be published by the CAA in its official record. (Civil Aviation Authority Regulations, SI 1991 No. 1672, Regulation 5). Pursuant to certain requirements in the Civil Aviation Act 1982 and the aforesaid regulations, the CAA publishes certain notices and other regulations in a regularly-updated system of publications known as the *Offical Record*. The *Official Record* is published through the CAA's publication division from where it can be obtained by subscription.

The *Official Record* is divided into a number of separate parts or series. In relation to air transport licensing, Series 1 forms an updated binder of regulations and notices particularly concerned with licence applications. Series 2 is a separate publication brought out every week or so giving notice of licence applications, hearings, decisions and miscellaneous items. It is Series 2 that is so important to persons who need to be promptly aware of licensing applications so that they may make an objection or representation if required. It is advisable for all airline operators to subscribe to at least Series 2 so as to protect their interests. Important information on other matters is also circulated through the *Offical Record*, such as tariff conditions and decisions.

Inspection

The CAA states in its *Official Record* (Section 11 as at July 1988) that it will also maintain in a prominent and accessible position a public notice board at its premises (CAA House) on which it will display the latest copies of its notices together with any other notices of interest to air operators. Copies of earlier air transport licensing notices and decisions are also available for consultation at the CAA's premises at CAA House.

Records of public hearings are available from the Licensing Section, ERG, and extracts from the UK DAT can be purchased from the Tariff Filing Unit, both of which are located at CAA House. Copies of decisions are available for purchase from the CAA's Printing and Publication Services at Cheltenham (and from the CAA Central Library at CAA House, London). The CAA Library is open to the public and apart from selling certain CAA publications has a useful reference library covering many aspects of aviation from commercial, statistical, technical, safety, training and other angles.

Practice note

The address of the CAA's publication division is Printing and Publication Services, Greville House, 37 Gratton Street, Cheltenham, Gloucester, GL50 2BN. Telephone: (0242) 235151.

Air Transport Users' Committee

Before leaving the CAA, mention should be made of the Air Transport Users' Committee or the AUC as it is commonly known. The AUC was set up by the CAA as an independent body to look after the interests of air transport users. In considering the general objectives of the CAA in performing its duties (see discussion on p. 24 above) it will be noted that one of its main statutory objectives is to 'further the reasonable interests of the users of air transport services'. It was with this in mind that the CAA established the Airline Users' Committee in 1973 as an independent body to make reports and recommendations on matters affecting air travellers. Its original terms of reference were:

> To assist the Civil Aviation Authority in its duties in safeguarding the interests of airline users and to investigate individual complaints against airlines where the person or body aggrieved has not been able to obtain satisfaction from the airline concerned.

It was, therefore, set up to protect the interests of consumers, i.e. the users of air services.

In 1978 its name was changed to the present Air Transport Users' Committee. Its revised terms of reference were to make reports and recommendations to the CAA for furthering the interests of air transport users. This was still to include the investigation of complaints against the suppliers of air transport services while giving it a more positive and active role in proposing changes for the future. The AUC is also to co-operate with any airport consultative committees which are charged by airport proprietors with furthering the interests of air transport users inside their airports.

The Committee itself is to consist of a minimum of 12 and a maximum of 20 members, including a chairman and deputy chairman. Appointments are normally for three years and are made by the CAA Board (on the joint advice of the Chairmen of the AUC and the CAA). In order to preserve its independent status no employee of the CAA may be a member of the AUC. Furthermore, members are to be appointed as individuals and not as delegates of any particular interest group and should represent as wide a range as possible of personal experience as users of air transport services.

The AUC has premises on the 2nd floor at Kingsway House, 103 Kingsway, London WC2B 6QX. The work of the Committee is supported by an annual grant from the CAA (see AUC, *Annual Report* for the year ended 30 September 1990).

Part II
Establishment of an Airline Operation

5 Airline Formation

Introduction

To start a new business venture is a daunting prospect whatever the nature of the project. Few individuals will have the variety of basic skills necessary today to operate a business ranging from the production, marketing and sale of the goods or services and managerial and administrative abilities to the appropriate financial and legal expertise. In a business of any size the entrepreneur will be required to organise and manage the various functions of the business while people with particular and appropriate skills and experience will be required to work within the business structure. In today's world it is also increasingly necessary to consult outside professional advisors such as accountants, lawyers and bankers. This general scenario becomes intensified when the new venture in prospect is an air transport operation.

The airline industry must be one of the most highly regulated areas of commerce. To operate civil aircraft for commercial purposes invokes a web of regulations affecting the 'operator', the 'aircraft' and the 'services' provided and we will be looking at some of the essential operational regulations in this part of the book. Any commercial airline operation involves a preparatory period of extensive planning. All too often enthusiastic would-be operators expect to put their ideas into practice within a few months of conception. Realistically this is a practical impossibility. The period between conception and commencement of commercial operations is long: it may last for many months, and, indeed, often several years. The first essential consideration is, therefore, the time factor.

Planning

Throughout the preparatory period any successful project will require detailed planning involving time-consuming attention and a high level of commitment. This should be recognised from the start: many promoters of airline projects are surprised and frequently disillusioned by the time taken and the details required not only to set up an airline operation but to comply with the regulatory requirements. Another important consideration is the financial commitment. Long before the first aircraft takes off on its inaugural flight considerable financial resources will have been spent in the planning stage. The time and advice of essential expert personnel and even the initial commitment to operational assets and facilities will be a costly investment, made before one penny of revenue is received.

These remarks are made not to discourage but to encourage proper awareness of the practical position so that appropriate plans can be made. Many new projects have failed literally to take off, so to speak, because their initiators have been ill-prepared and ill-informed. For the newcomer, it is a difficult industry to enter; for an established airline operator at the start of a new operation, the task is by no means so daunting: it is essentially a task of plodding through established familiar procedures. For the uninitiated a start-up situation is full of potential problems. These can, however, particularly with foresight, be overcome without too much difficulty. In this chapter we will explore some of the essential steps which have to be taken in setting up a commercial airline operation.

Operating body

First and foremost such an undertaking will be a business and will entail all the usual considerations inherent in that. We, however, will look at those considerations specifically related to an airline operation. Suffice it to say that, in general terms, the prospective operator will require a vehicle for the proposed airline operation. Whilst the term 'operator' generally referred to in many regulations can technically include an individual, partnership or body corporate (i.e. a company) it is the limited company that is almost exclusively chosen. There are, of course, a number of reasons which make a corporate structure most appropriate, including the complexity and sophistication of an airline operation, the need to attract substantial investment and the need to limit the risk of potentially large liabilities.

A new company incorporated under the companies acts will need to be acquired or formed from scratch. Its own rules and constitution as laid down in its Memorandum and Articles of Association will require adapting to the requirements of the proposed airline. In particular, consideration at this early stage should be given to the financial structure to provide for the required investment. It will be seen from the discussions of 'financial fitness' requirements (p. 48) that there is a statutory requirement for the CAA to be satisfied that an operator is adequately funded before granting an air transport licence.

Another general consideration to be borne in mind from inception is the obtaining of appropriate key personnel and shareholders. In particular, as we will review in more detail below, air operator certificates and air transport licences can be refused if the relevant people within the operation are not deemed fit by the CAA. There is also a further more fundamental consideration to be taken into account from the outset: the nationality of the operator. In the case of a body corporate this primarily means shareholders. Statutory regulations require the CAA to ensure that an applicant for an air transport licence is a UK national or being a body corporate is controlled by UK nationals. Again, these requirements will be reviewed further when looking at the licensing regulations in Chapter 6.

From the very beginning, therefore, there are a number of considerations affecting any proposals for an air transport operation which relate exclusively to the air transport business and provide an additional and significant regulatory burden not existing in other areas of commercial activity. In the following chapters we will review some key operational regulations. Whilst most of these relate to the ongoing obligations of an airline operator they demand the specific attention of the prospective operator who will also need to apply for and satisfy the authorities that it should be granted certain licences, certificates and other approvals before it can commence operations. There are three key areas of regulatory control.

1. Operator

The operator being an individual or a body corporate, will in the first place have to obtain an air operator's certificate (AOC). From this many regulatory obligations will ensue.

2. Air transport licensing

As the backbone of the statutory control of economic regulation, the CAA is empowered to grant licences to applicants who qualify in respect of air transport services on specified routes. Route licensing is, therefore, a vital part of an airline's operation. In effect, the licensing system through its economic control is ultimately responsible for the success or failure of commercial air transport. Whether or not an operator is licensed on the routes it wants dictates whether or not it can supply those services and carry on in that business. For this reason route licences are valuable assets and are often aggressively defended against applicants for licences on competing routes.

3. Aircraft

There are many different areas of regulation affecting the operation of civil aircraft. First, a prospective operator must obtain a certificate of airworthiness in respect of each aircraft. This initial certification is backed by extensive continuing obligations particularly in the area of aircraft checks and maintenance.

While airworthiness regulation relates primarily and directly to operational safety, there is another important area of regulation which deals with the administrative organisation of aircraft existence. This is the registration which involves, first, the registration of all UK civil aircraft, details of each recorded in a central register, and to each aircraft the conferring of an individual registration number; and second, the registration of aircraft mortgages and charges affecting individual aircraft which are also recorded in a central register maintained by the Civil Aviation Authority.

Another area of importance to the operators of aircraft are the incumbencies that may affect their aircraft. Certain authorities have rights to retain and even sell aircraft for non-payment of certain charges or taxes. For example, the relevant airport authority has statutory rights under the Civil Aviation Act 1982 to retain aircraft for non-payment of airport charges and similar provisions relate to authorities which provide navigational services.

These three areas are fundamental to the operator in its capacity as an airline manager, aircraft operator and seller of air transport services. These areas will be considered in more detail below. It is perhaps worth noting that in defining the requirements for each certificate, licence or approval the basic definitions of applicability relate to aircraft and their nationality as a starting point; namely 'aircraft registered in the United Kingdom'. After examining these three basic regulatory areas to which every prospective operator will need to prove compliance and, subsequently, be granted appropriate approvals, we will turn to other areas which regulate and affect the industry in general.

6 Air Transport Licensing: The Role of the CAA

As we have seen the law and regulation of civil aviation is primarily the role of the Civil Aviation Authority (CAA) and nowhere is this more apparent than in the area of economic regulation. As the body responsible for granting, revoking, suspending or varying an air transport licence it has certain essential statutory obligations when performing these vital functions. These duties are primarily laid down in the Civil Aviation Act 1982 (referred to as 'the Act').

1. The licencing of air transport is a function conferred on the CAA by or under the Act (Section 3).
2. The Act states certain general objectives for the CAA in performing its duties. In particular, it is the duty of the CAA to perform the functions conferred on it in the manner which it considers is best calculated:

   (a) to secure that British airlines provide air transport services which satisfy all substantial categories of public demand (so far as British airlines may reasonably be expected to provide such services) at the lowest charges consistent with a high standard of safety in operating the services and an economic return to efficient operators on the sums invested in providing the services and with securing the sound development of the civil air transport industry in the United Kingdom; and
   (b) to further the reasonable interests of users of air transport services

These objectives are fundamental (and have therefore been quoted verbatim from the Act) (Section 4). The aim is to provide a wide range of air transport services while protecting the interests of both the airlines and their users. It is essentially an attempt to balance two different interests: to ensure that the consumer enjoys the lowest charges consistent with a high standard of safety whilst ensuring that the efficient operator enjoys an economic return on its investment.

A British airline for these purposes is defined as an undertaking having power to provide air transport services and appearing to the CAA to have its principal place of business in the UK, the Channel Islands or the Isle of Man and be controlled by persons who are either UK nationals or are for the time being approved by the Secretary of State. These provisions (Sections 3 and 4) confer the basic function of air transport licensing and lay down the general objectives within the context of the airline industry.

The Act also provides specific provisions for the carrying out of the air transport licensing function. In particular, the role of the CAA is clarified by its various duties and obligations. Before looking in more detail in subsequent chapters at the licensing procedures as they affect the various parties we have to consider here the role of the CAA and the general criteria by which it must exercise that role. We are not therefore considering the rights to a licence or the methods and procedures by which to obtain and alter such a licence, but the general principles of policy laid down in the Act which should be borne in mind before the former are considered. In exercising its licensing function the CAA must consider the following criteria.

Nationality

The inherent nature of aviation poses questions of nationality, jurisdiction and conflict of laws, and thus the issue of nationality must be addressed, defined and restricted. We have already seen how some of the fundamental questions were considered, and in part addressed, by the Chicago Convention of 1944. Clearly, in route licensing, the lynch-pin of economic regulation, an independent state must define exactly who in terms of operators will be within their jurisdictional net of regulations. In the UK the 1982 Act (Section 65 (3)) dictates the definition of a UK national for licensing purposes (which in effect covers civil air transportation for reward).

The CAA must not grant an air transport licence to any applicant unless it is satisfied that that person is a UK national or, in the case of a body corporate, (essentially a company as opposed to an individual or group of individuals) is a body which is incorporated under the law of any part of the UK or the law of a relevant overseas territory or an associated state and is controlled by UK nationals. The CAA has no power to exceed or exempt this restriction. The Secretary of State, however, has the ultimate power, acting on behalf of the government, to intervene and give consent to the grant of licences in circumstances where the CAA is not itself satisfied that an applicant is within the above criteria. Furthermore, to ensure the Secretary of State has the ultimate say over questions of nationality the CAA must notify it where it proposes to refuse to grant a licence purely on nationality grounds and to postpone its decision until the Secretary of State's consent is given or refused.

The CAA has formulated its own criteria for control, particularly relevant in the case of body corporates. The CAA has (over the years) stated that it requires evidence of substantial ownership and effective control in the hands of United Kingdom nationals. It has always refused to specify an exact measure for control. For example, when considering share capital in a company it does not specify an exact percentage which must be owned by UK nationals. Some countries, for example the USA, state a figure, such as 75 per cent of the share capital, which must be owned and controlled by nationals of that country. In the UK it is left to the discretion of the CAA thus allowing it the opportunity to look more thoroughly at each situation.

The CAA argues that, for example, the question of actual control may not merely be a matter of who owns most of the shares. It considers who owns the balance, how it is split, for example, amongst different shareholders and what other influences they may be able to exert. For instance, if, say, 65 per cent of a company is owned by one UK national who also controls the board, and the remaining shares are owned by a mixture of non-nationals who as individuals have no other interest in the company other than to protect their investment, it is arguable that 65 per cent could be sufficiently acceptable. If, however, the 65 per cent was owned by a mixture of individual UK nationals who had no other involvement in the company and 35 per cent was owned by a non-national leasing organisation that has several seats on the board and has leased aircraft to the company with rights to appoint a receiver (in certain situations of default under the leases) it could have a contrasting effect. This example would clearly be seen as quite different from the previous one and presumably the CAA would be far less likely to see this as acceptable.

The CAA, therefore, will investigate quite carefully the true position of control and is able to require answers to quite penetrating questions in order to ascertain whether it is satisfied, in accordance with its statutory obligations, that the applicant is a UK national.

It should also be remembered that this is a continuing criteria. Section 66 also provides when dealing with the suspension and revocation of air transport licences, that the CAA must, as it thinks appropriate, revoke, suspend or vary air transport licences in certain situations. These include circumstances where the CAA is no longer satisfied that the holder (having regard to various considerations) is fit or has adequate resources and where the CAA has reason to believe that the holder of the licence is not a United Kingdom national as defined. In the latter case it is the duty of the CAA to inform the Secretary of State accordingly, and if he so directs, revoke the licence.

The whole issue of nationality is one, of course, that is likely to be completely reviewed in relation to the position of the UK in the European Economic Community (EEC). Many potential airline entrepreneurs are even now surprised to learn that the nationality requirements remain in a form that, prima facie, resists joint operations with nationals from other EEC countries. For instance, a proposal for a new commuter airline operating out of the south of England to points around Europe with investors from various European countries each represented on the board, though in true Common Market spirit, could have problems. Under the present statutory regulations it would be unlikely, if control were shared amongst Common Market partners, for it to pass the UK national test as far as the CAA is concerned. However, it has always been open to the Secretary of State to consent to an applicant who would otherwise be rejected purely on nationality grounds. In view of Britain's membership of the Common Market, one may expect a more sympathetic approach towards an application involving EEC 'nationals'.

As standardisation and further European co-operation develops towards the first of January 1993 and beyond, no doubt a joint approach will be adopted towards the issue of nationality, probably along the lines of an EEC definition which will apply to any applicant for an air transport licence in the EEC. This would obviously encourage European investment and joint venture operations and greatly enhance advances towards creating a true European Market.

Fitness

Another essential criterion of the licensing system is that any prospective applicant must prove it is fit to be the holder of the licence for which it has applied. This sort of legislation is again common in most nations and is generally considered vital to protect the interests of the travelling public both commercially and in terms of safety.

The fitness qualification deals with two distinct areas (see Section 65). First, that the applicant is a fit person to operate aircraft under the licence. In considering the applicant's suitability the CAA must have specific regard to his and his employee's experience in the field of aviation and their past activities generally. Where the applicant is a body corporate the CAA must have regard to the experience and past activities of the persons appearing to the CAA to control that body. Second, the CAA must consider the financial suitability of the applicant. In particular, it must be satisfied that the resources of the applicant and the financial arrangements made are adequate for discharging the actual and potential obligations in respect of the business activities in which the applicant is engaged (if any) and in which it may be expected to engage if it is granted the licence (which the CAA thinks should be granted in pursuance of the application).

Again, the CAA pays particular attention to these vital areas and a department within the CAA deals solely with the assessment of the financial fitness of an applicant. Investigation is particularly thorough in the case of a new applicant especially if it involves a completely new airline. It is left totally to the discretion of the CAA as to what tests it applies and how it interprets the statutory criteria which it is bound to apply.

Before licences are granted or a hearing commenced, an applicant must normally have satisfied the financial fitness regulations of the application. In the case of totally new applicants this can take time, perhaps several months or more, and consequently can delay any hearing of the application(s). The applicant must produce a detailed business plan with complete financial forecasts for a minimum of the first two years of operation. These must show all income and expenditure over the period on a monthly basis. For an established concern a considerable amount of historical information is required in the form of audited accounts and recent management accounts. The ERG (Economic Regulation Group) Airline Finances Division of the CAA normally specifies in writing the precise information that is required. Questions and other information may be discussed on an informal basis but then a hearing (in the nature of a formal meeting) is normally held, attended by the applicant. This is chaired by a CAA board member with authority for airline finances. Following the hearing the applicant will be informed as to whether the CAA considers the financial arrangements and proposed resources are satisfactory or what might be required to enable it to be so satisfied.

Experience shows that, particularly in the case of new airline operators, an applicant can encounter delays in obtaining finance or, at least, a commitment to provide funds before licences are granted. There is always this problem of funding a start-up airline. Apart from the historic reluctance of normal institutional sources of finance to invest in airline projects, investors are naturally reluctant to commit funds to what is little more than a shell company before licences are granted and trading, in terms of flying services, commences. However, without some commitment, or at least funding conditional on licences being granted, the CAA naturally does not wish to devote the time and trouble necessary for a hearing and its preparation if the appropriate finance is not forthcoming at the end of the day.

Furthermore, at many times in the 1980s, the number of applicants waiting for hearing dates was high and delays long. The number could have been considerably and unecessarily lengthened were every potential applicant included. Applicants often underestimate the time-scale in which they will be able to commence operations. They need to recognise that the preparation of business plans (with all the information and planning that will entail), meeting the funding requirements and preparing for a possible hearing will take a considerable amount of time.

UK airline considerations

The next general principle for the CAA to consider is the potential impact on existing services of granting a licence to a particular applicant. It has a statutory duty when considering whether to grant a licence to have regard to the effect on existing air transport services provided by British airlines of authorising any new services the applicant proposes to provide under the licence (Section 68 (2)). In any case where these existing services are similar (in terms of route) to the proposed services, or where two or more applicants have applied for licences to operate similar services, the CAA must have regard in particular to any benefits which may arise from enabling two or more airlines to provide the services in question.

In considering such issues the CAA will, of course, need to bear in mind the general statutory objectives as stated above (p. 45). For example, is existing and potential demand likely to be sufficient to ensure that with the present and the proposed capacity offered by the new applicant, at least one or more operators on the route will still be able to operate profitably so that they receive an economic return from efficient operations on sums invested (Section 4)? Whilst trying to protect those interests, not only for the sake of airlines, but also to try and create stability within the market place, the CAA has to bear in mind the interests of the consumer such as increased capacity, choice and more competition which encourages lower charges (again within the general statutory objectives).

The policy of the CAA as expressed in its policy statements (e.g. CAP 539 Statement of policies on air transport licencing – June 1988) has developed throughout the 1980s towards increasing competition. It is true to say that over the years there have been considerable changes in policy and the attitude of decision-makers. From the pre-1970s establishment position of a single monopolistic flag carrier (still the view and position in many countries today) to the introduction of competition on a limited basis in the late seventies, we have seen the active encouragement of competition in the more liberal 1980s. The old schools of thought which believed in single designation, i.e. only one airline per route, have been largely replaced by a multi-designation policy. This has had the effect of gradually changing the fundamental criteria of decision-makers towards the granting of licences. For instance, where previously the onus was very much on the applicant to prove why it should be granted a licence, there is more recently a tendency to reverse this more negative approach to one where an objector must show good reason why the applicant should not be granted the licence. This is not perhaps an acknowledged technical change but is a subtle, slowly developed change of emphasis which has resulted in a more liberal approach.

Bilateral considerations

Having considered the impact on the home market, the CAA also has a statutory duty to consider the international implications. The CAA has, therefore, to perform its air transport licensing functions in the manner which it considers is best calculated to ensure that British airlines compete as effectively as possible with other airlines in providing air transport services on international routes. In performing these duties it must have regard to any advice received from the Secretary of State with respect to the likely outcome of negotiations with the government of any other country or territory for the purpose of securing any right required for the operation by a British airline of any air transport services outside the United Kingdom. The CAA must also have regard to the need to secure the most effective use of any airport within the UK (Section 68 (1)).

This essentially ensures that bilateral considerations, in this instance the international traffic right position, are taken into consideration when granting air transport licences. Potentially, of course, the licensing of the second carrier on a route could prejudice an existing carrier and maybe even the national interest. For example, with the traditional type of bilateral agreements, often based on single designation (in other words each country allowed one carrier so that there is a closed total of two on the route), the introduction of a second carrier by one country could cause problems. In the first place the other country might refuse to allow the designation of a second carrier to its bilateral partner, or it might negotiate either to designate a second carrier itself (if that were possible) or accept the second carrier for its bilateral partner on the basis that the overall capacity (in terms of flights or seats) of that bilateral partner must not exceed that which it provides.

Whatever the outcome of the resulting negotiations, it is a fair certainty that there will be a price to be paid for any new benefit gained by the requesting partner (except in cases where liberal agreements exist). The last option above, means that the existing home carrier plus the new home carrier must, in effect, share the capacity previously provided totally by the existing carrier who is going to strongly resist this option. Unfortunately, on traditional diplomatic lines, any request for a change in a bilateral agreement by one side, is inevitably going to result in the other trying to negotiate an equally good benefit for itself. There is, therefore, normally a price to be paid for introducing any new carrier on an international route. This bilateral price must be assessed by the CAA on the written advice of the Department of Transport (see bilateral issues at hearings, p. 69).

Environmental issues

Particular attention is to be paid to environmental issues with increasing awareness of, and opposition to noise and pollution caused by aircraft. Section 68(3) places a specific statutory duty on the CAA to have regard to the need to minimise so far as is reasonably practicable any adverse effects on the environment and disturbance to the public from noise, vibration, atmospheric pollution or any other cause attributable to the use of aircraft for the purpose of civil aviation. On the face of it, this is widely drafted and far-reaching in its application. This is stated, however, to be subject to Section 4 (see p. 46) and Section 68(1) and (2) (see pages 50 and 49).

This qualification avoids to some extent a fundamental conflict of interest between the operators of aircraft and environmentalists. Aircraft do make a noise and can pollute the atmosphere. The devil's advocate might otherwise ask how the CAA is to carry on performing its air transport licensing functions and minimise as far as is reasonably practicable such effects as 'any disturbance to the public'. In the real world, of course, the CAA as the licensing authority is limited in what it can actually do to minimise such effects. It is its job to license airlines and environmental issues have to be considered within the context of that function. Clearly, when an environmental problem within the statutory definition emerges that is particularly significant its presence could affect the outcome of a licence application.

Perhaps the best example in recent years of environmental issues impacting on a licensing decision is that of the London City Airport, first opened in 1987. Initially, two operators were licensed to fly services from the airport, a stolport (i.e. an airport with a short take off and landing capability), situated in the middle of a built-up area in London's docklands. The CAA thought fit, particularly in light of the concerns of noise protestors, to grant the licences subject to a number of conditions to limit the impact of aircraft noise on the surrounding area. For example, restrictions were placed on the type of aircraft to be used and the number and hours of aircraft movements.

Liberalisation

Very much a part of the new liberal approach of the early 1980s in working towards an increasingly competitive market is the expressed conclusion in the 1982 Act that the CAA has a duty basically to minimalise restrictions that it may impose (Section 68 (4)). The Act states in a somewhat long-winded fashion that in addition to the duties with respect to particular matters imposed on the CAA by preceding provisions of that section, it shall be the duty of the CAA to perform its air transport licensing functions in the manner which it considers is best calculated to impose on the civil aviation transport industry of the United Kingdom and on the services it provides for uses of air transport services *the minimum restrictions* consistent with the performance by the CAA of its various statutory duties (specified as those under Sections 4, 65 and 66).

This is a clear indication that within its statutory restrictions the CAA should be as liberal as it is able when exercising its duties and functions. Within these statutory obligations and conditions, the CAA has a wide discretion, and, in general, an air transport licence may contain such terms as the CAA thinks fit (Sections 65 (5) and 68 (4)).

7 Air Transport Licensing: the UK System

The system of route licensing in the UK is a fundamental part of the regulation of the airline industry. It is mirrored by similar systems in many other countries. When looking at the licensing regulations in the UK we must consider three sources:

(1) Originating legislation – the Civil Aviation Act 1982.
(2) Specific regulations made pursuant to that Act – the Civil Aviation Authority Regulations 1991 (SI 1991 No. 1672) and subsequent amendments.
(3) Civil Aviation Authority publications – the *Official Record*, Air Transport Licensing, Series 1.

The Civil Aviation Act 1982

Part III of the Act deals with the regulation of civil aviation and, in particular, route licensing in Sections 64 to 71. Any actual or prospective licensee should be familiar with these provisions.

The essential requirements for an air transport licence are laid down in Section 64 (1). No aircraft is to be used for the carriage for reward of passengers or cargo on a flight to which the subsection applies unless the operator of the aircraft has an air transport licence granted by the CAA authorising it to operate aircraft on such flights and it is complying with the terms of the said licence.

Again, as we shall see with the air operator's certificate (see Chapter 14), the requirement is related to aircraft registered in the UK. Section 64 (2) clarifies those flights which are subject to this section:

> any flight in any part of the world by an aircraft registered in the United Kingdom and to any flight beginning or ending in the United Kingdom by an aircraft registered in a relevant overseas territory or an associated state.

There are a number of specified exceptions concerning flights for which a licence is not required. These include those flights for which the CAA is the operator or has specifically excluded from these requirements in accordance with the procedure stated in Section 64. For these exceptions reference should also be made to the CAA's *Official Record*, Series 1, (and amendments notified in the Series 2, publications) where it provides from time to time for certain categories of flight that are exempt from the need for a licence. For example, these include certain categories of flights for specific purposes. The list of these is extensive and should be consulted before embarking on an application.

The Civil Aviation Authority Regulations 1991

These regulations made pursuant to the 1982 Act are produced, updated and amended where necessary, by statutory instruments. They provide the more detailed regulations whereas the Act itself tends to lay down the general principles.

The *Official Record*

The CAA under cover of its publication, the *Official Record*, Series 1 regularly publishes notices on matters affecting air transport licensing. This provides updating information on such matters as categorising the various classes of air transport licence, those flights that are exempt from the requirements for an air transport licence, and standard conditions attached to air transport licences and tariff conditions.

One of the CAA's most important functions is, in Series 2, to publish on a frequent basis (normally every week or so) licensing notices relating to the publication of applications for air transport licences and notification of hearings and decisions thereof (as well as other important notices). This publication is essential reading for airlines and other interested parties who may wish to object to particular air transport licence applications, variations, revocations or otherwise. Bearing in mind (as we shall see in Chapter 9 on objections and representations) that there is a time limit for submitting an objection, it is essential that the information on which the objection would be based is publicly available.

8 Air Transport Licence Applications

Application for the grant of an air transport licence

Section 65 (1) of the Civil Aviation Act 1982 states that an application for the grant of an air transport licence must be made in writing to the CAA and must contain such particulars as the CAA may specify in a notice published in the prescribed manner. It should be noted that we are talking here merely about the granting of licences not the revocation, suspension or variation of licences which are covered under separate provisions.

Pursuant to its statutory rights and duties (see Chapter 3) the CAA publishes notification of certain regulations and requirements in respect of air transport licensing in its *Official Record*, Series 2. Series 1 of that publication states in full a list of information required pursuant to Section 65 (1) which the applicant should include in its application for the grant of a licence from the CAA. This is reflected in the questions asked on the appropriate form. The list of information in general includes such matters as the following:

(1) particulars of the applicant (e.g. name, address, nationality);
(2) previous aircraft and operations;
(3) financial resources and arrangements;
(4) aircraft and organisation for proposed operations;
(5) liability provisions (e.g. insurances etc.);
(6) details of licence applied for (e.g. need and demand for operations, class of licence, tariffs, routes, aircraft, etc.).

(See *Official Record*, Series 1, Section 3, p. 15, November 1990)

Application forms

Application should be made on the appropriate form. All forms can be obtained from the Air Transport Licensing Section, (ERG Division), Civil Aviation Authority, 45–59 Kingsway, London WC2B 6TE.

An applicant has to submit either one or two specific forms. The first form is that which relates to the actual licence applied for and the relevant form corresponds to the class and type of licence required. For example, form 101 (A) is for a class one (A or B) scheduled licence. Section 3 of the *Official Record*, Series 1, lists the appropriate form for each particular licence. A second form has to be completed and submitted where it is a first application for a licence. This form (reference ATL 5) is essentially to provide information about the applicant. For example, in the case of a company, questions are asked in relation to share capital and shareholders; parent, subsidiary or associated companies; control; directors, managers and staff.

The original of the first, and where appropriate, the second form should be sent together with four copies of each to the CAA (at the address stated above), accompanied by any application fee. The CAA has a right to levy charges for the carrying out of its statutory functions under Section 11 of the Civil Aviation Act 1982.

Application for the revocation suspension and variation of an air transport licence

For such applications (not being for the *grant* of a licence) there is a similar procedure but with a few noticeable differences. The right of application is again given by virtue of the Civil Aviation Act 1982 under Section 66.

An application for the revocation, suspension or variation of an air transport licence may be made to the CAA at any time by 'a person of a prescribed description' (Section 66 (1)). The Act does not define such persons but by use of the word 'prescribed' means as further described by regulations made by the Secretary of State. Thus, clarification is found in the Civil Aviation Authority Regulations 1991 under Regulation 16(6). Any one of a list of prescribed persons can apply to the CAA for the variation, suspension or revocation of an air transport licence, but, except as provided, no person may apply for the variation of a schedule of terms (see Regulation 18). The list of prescribed people (specified as those in Regulation 25(1)(b) to (d) is essentially the following:

(b) the holder of any air transport licence;
(c) the holder of any air operator's certificate granted under an Air Navigation Order;
(d) the holder of any aerodrome licence granted under an Air Navigation Order.

All applications (whether for the grant, revocation, suspension or variation of a licence) should be made on the appropriate forms which can be obtained from the licensing section of the CAA (see p. 55 for address).

Once completed, the application forms (together with four copies of each form) should be submitted to the CAA accompanied by any appropriate charge. The CAA has the right to levy charges for the carrying out of its statutory functions (see Section 11 of the Civil Aviation Act 1982) and Regulation 16(1) provides that the CAA may refuse to consider an application unless the applicable charge accompanies the application.

Notice requiremets

In general the CAA may refuse to consider an application for the grant, revocation, suspension or variation of a licence unless (subject to regulation 24) it has been served on the CAA not less than six months before the beginning of the period for which the licence or otherwise is proposed to be in effect (see Regulation 16(1)). The CAA, however, states (in its *Official Record*, Series 1, Section 3 (6)) that it has a discretion to accept applications for the grant of a licence at less than six months notice and will 'normally' do so subject to an absolute minimum of ten working-days' notice. This is merely discretionary and cannot be relied on. Applicants should, therefore, always be prepared to plan for a notice period of at least six months unless perhaps there is some justifiable urgency in respect of their application.

9 Objections and Representations

It is a vital part of the UK air transport licensing system that interested parties have the right to object to any licence application. Compared with most forms of business and commercial behaviour this is actually a strange phenomenon. For what is really happening is that other operators in the market place are able to oppose the licensing of actual or potential competitors to carry out the operations they wish to commence or continue. Whilst the right of opposition is supposed to be exercised in a justifiable manner to serve the better good of the industry and the consumer, it is perhaps extraordinary (if not unique) that competitors are able to influence whether or not another party will be able to carry on business alongside them. The reason for this goes to the root of why such control and regulation is necessary in aviation. Without regulation for the purposes of public safety and protection, operational organisation and international relations would result in chaos in the sky. Economic regulation has been accepted, therefore, as a vital part of governmental control and the right to object is a fundamental part of that regulation.

Who may object

The Civil Regulation Authority Regulations 1991 (SI 1991 No. 1672 as amended) state categorically that any person may serve on the CAA an objection to or representation about an application or proposal for the grant, revocation or variation (other than a provisional variation) of an air transport licence (see Regulation 20(1)). An objector may not, however, have the right to be heard (see below).

Time and manner

An objection or representation should be made in writing and submitted to the CAA's licensing section, ERG Division, Room T508, 45–59 Kingsway, London WC2B 6TE. Normally each objection or representation should relate to one application only (see CAA, *Official Record*, Air Transport Licensing, Series 1 Section 7 as published November 1990) and it is vital that the person making the objection or representation states whether or not they wish to be heard by the Authority. If they do not do so, they forfeit their right to be heard under Regulation 25 (1). A person who does not have a right to be heard will not be able to present their case to the CAA (see Regulation 25).

Regulation 25 (1) specifies the parties that have a right to be heard before any decision to grant, refuse to grant, revoke or suspend or vary (other than provisionally) an air transport licence is made (see Chapter 10 on hearings). That right is conditional on a number of points (see Regulation 25 for special cases). Two of these conditions are particularly significant here. First, no person (unless they are the applicant or holder of the licence to which the decision will relate) will have the right to be heard unless they have served an objection or representation in accordance with the regulations and second, if they are a person specified in 25 (1) (a), (b), (c) or (d) of the regulations (namely, the applicant, holder of an air tansport licence or any air transport operator's certificate granted under an Air Navigation Order or the holder of an aerodrome licence granted under an Air Navigation Order) unless they have stated in their objection that they wish to be heard. Thus, in accordance with the regulations, where an objector (or person making a representation) has served an objection on time to the CAA they will technically not be entitled to be heard at the hearing, unless they have stated that they wish to be so heard.

In conclusion, therefore, an objector or person making a representation must do so within the time and manner stated and include a statement as to whether they wish to be heard. There are stringent time limits within which any such objection or representation should be served on the CAA. The time limit is stated in the notification of the application in the *Official Record* which is itself dated. That date is, in effect, the date of publication and there is normally a period of 21 days from that date during which the service of the objection must take place.

The regulations state the bands within which the CAA may specify time limits when publishing an application or proposal. The normal rule is that the time limit must be not more than 21 days nor less than 7 days. Where however, in special cases, an application or proposal is not published and the person wishing to serve an objection or representation has been notified by the CAA that the application or proposal has been made and will not be published, the time for service is within three working days of being so notified (see Regulation 20).

The CAA must not specify a period of less than 21 days for the service of objections and representations unless it is satisfied that it is desirable to do so for reasons only of urgency or in special circumstances relating to the revocation, suspension or variation of licences (see Regulation 17) in accordance with the direction of the Secretary of State or to vary a licence for the sole reason that there is a need to allocate scarce capacity (see Regulation 20).

There is a further exception relating to an environmental application or proposal where the CAA must specify a limit of 42 days from that of publication (pursuant to Regulation 19) for the service of objections or representations on the grounds of noise, vibration, pollution or other disturbance. In practice, a prospective objector or person wishing to make a representation should check time limits specified by the CAA in relation to the particular application.

Service on other parties

Where the person making the objection or representation is the holder of an air transport licence they must, within 24 hours of serving it on the CAA, serve a copy on certain other persons. Those persons are the applicant, any other person who is a holder of the licence to which the application or proposal relates or any other person which the Authority is obliged to consult (under Regulation 21 in relation to flights to, from or within the Channel Islands, the Isle of Man, or Gibraltar) (see Regulation 20 (3)).

Where the person making the objection or representation is not the holder of an air transport licence the CAA shall serve on the above stated persons a copy of the objection or representation within seven days of it being served on the CAA, indicating whether the person making the said objection or representation wishes to be heard (pursuant to Regulation 25). There is also a provision entitling the person making the objection or representation to be served by the applicant with a copy of the application within three working days of being requested to do so.

Special cases

It must be remembered that the above regulations relate to the majority of cases. There are a number of exceptions or special cases for which different regulations may apply, for example, the application to vary a schedule of terms or environmental cases. The actual regulations (both made under statute and by the CAA) should always be consulted. In any particular case, one can always consult the CAA for advice.

Practice guide

1. All persons who may wish to make an objection or representation in respect of any air transport licence application should ensure that they receive immediate notification of such applications. The CAA publishes notification of applications in its *Official Record*, Series 2, Air transport licensing notices. Subscription to that publication can be arranged by contacting the CAA, Printing and Publication Services, Greville House, 37 Gratton Road, Cheltenham, Gloucestershire GL50 2BN, Tel: (0242) 235151.
2. If a person wishes to object or make a representation to the CAA this must be made strictly within the time limit. This is normally (but not always) within 21 days following the date of publication of the application or

proposal. (The appropriate dates are stated on the CAA notification of the application in its *Official Record* – see above).

3. An objection or representation must be in writing and sent to the CAA at the address stated on the publication of the application. This does not require the completion of a particular form and can be achieved by a simple letter. Any reasons for the objection or representation do not have to be given at this stage but should be submitted by way of a written submission as provided by the CAA (see Chapter 10 on the 'preparation for a hearing'). The letter *must* state if the person making the objection or representation wishes to be heard. A simple and general example of a form of letter is provided in Figure 9.1 and can be adapted for use with the appropriate changes.

4. It is essential that a copy of the objection or representation is served on other relevant parties within 24 hours of the original being served on the CAA. Figure 9.2 provides a simple form of letter to accompany the copy. The relevant parties are the applicant, any other person who is the holder of the licence to which it relates and any body which the CAA is obliged under Regulation 21 to consult. In general, this affects flights to, from or within the Channel Islands, the Isle of Man, and Gibraltar. Section 7 of the *Official Record*, Series 1, provides the names and addresses of the bodies that should be served in specific cases.

Figure 9.1 Specimen letter of objection or representation to the CAA

From: Objector

To: Civil Aviation Authority
        ERG Division
        Room T508
        CAA House
        45–59 Kingsway
        London
        WC2B 6TE
        [Or other address specified in Official Record]

Dear Sirs

[Reference: Applicant .....
              Application No(s) .....
              Routes/sectors .....]

We hereby give notice of [this company's] objection to the above application published in the *Official Record*, dated [as dated on notice in *Official Record*], and of our intention to be heard [represented] at any hearing thereof.

We confirm that we have sent a copy of this letter to [applicant] at [applicant's address].

Please acknowledge receipt hereof and confirm that we are not required to serve notification on any other person.

Yours faithfully,

[Objector]

Figure 9.2 Specimen letter to relevant parties to accompany copy of objection or representation to the CAA

From: Objector

To: [Applicant]

Dear Sirs,

[Reference: as previous letter (Figure 9.1)]

We enclose a copy of the letter sent by us to the Civil Aviation Authority today, giving notice of our objection to the above application.

Please acknowledge receipt hereof [and supply us with a copy of the application].

Yours faithfully,

[Objector]

10 Preparation for a Hearing

Once the hearing date for a licence application has been set the parties must commence careful and systematic preparations for the hearing. There are various points to consider but the immediate concern is that of evidence to be served by written submission before the hearing commences.

This requires a considerable amount of work and will involve the input of all those persons who can contribute to the case of the party be it an applicant or objector. Past experience has shown that the presentation of a successful case needs to be run as an organised campaign with a team of essential people. If lawyers are instructed it is usually led by the solicitor, otherwise a suitably-experienced person within the party's organisation should be appointed 'campaign leader'.

The task of the team is to construct its party's case by clarifying its goals, collating and assembling evidence, and finally producing a case in two forms: first, in writing by way of submissions, and second, an oral presentation at the hearing itself. From the announcement of the hearing date (normally at least two months before the hearing) the parties must proceed on a regulated time-scale. This relates particularly to the service of written submissions commencing with the service of the applicant's Evidence-in-Chief.

Written submissions

The CAA provides the essential timetable for submissions in its *Official Record*, Series 1, Section 9 headed 'Hearings and Decisions of the Authority'. Applicants must first produce their written submissions and are expected to deliver them at least 25 working days (basically five weeks) before the start of the hearing. Failure to do so is likely to be regarded as a compelling reason why the hearing should be put back. This is unlikely to be in the interests of the applicant who is, therefore, wise to deliver its submission on time. This submission, being a written document presenting the applicant's case, is sometimes referred to in more legalistic terminology as the Evidence-in-Chief.

There is sometimes more than one applicant where, for instance, several applicants apply to be licensed on a new route. In such a case where there are competing applicants and one delivers its submission on time but another does not, the CAA will normally hear the former's application on the due date and set a new, later date, for hearing the latter.

It is then the turn of the objectors to deliver their written submissions which they are expected to do at least ten working days (normally two complete weeks) before the start of the hearing. This means they will have had up to 15 working days (approximately three weeks) to see and consider the case of the applicant or applicants. In legal terms the objectors' evidence is sometimes referred to as Evidence-in-Rebuttal as essentially it rebuts the Evidence-in-Chief of the party bringing the action, known here as the applicant. If the objectors fail to deliver their evidence within the time limit the CAA will consult the applicant as to whether it wishes the hearing to be postponed or to proceed as arranged. This protects the interests of the applicant and is necessary to ensure that it is not prejudiced by the behaviour of objectors. Otherwise in practice, objectors could try and cause indefinite postponement of the hearing by delaying the service of their submissions.

Service

Regulation 4 of the Civil Aviation Authority Regulations 1991 provides for the proper service of anything required to be served under the regulations or under Sections 66 (4) or 84 (1) of the 1982 Act. Anything requiring to be served must be set out in writing and served to the appropriate person by one of the following means:

(1) by delivering it to that person;
(2) by leaving it at that person's proper address;
(3) by sending it by post to that person at that address; or
(4) by sending it to that person at that address by telex or other similar means which produce a document containing a text of the communication in which event the document shall be regarded as served when it is received.

Where the person is a body corporate (e.g. a registered company) the document may be served upon the secretary of that body. The proper address in the case of a body corporate shall be its registered or principal office and in any other case the last known address of the person to be served.

Form and content

The applicant's submission is essentially to outline its case, so as to demonstrate why its application should be granted, by a detailed analysis of its suitability, for example, any strategy or concept behind the proposed operations; the need and demand for such services, and the benefits to the applicant and the consumer without an unacceptably adverse impact on any competitor.

There are no fixed rules about the form and content of submissions although certain basic information is clearly required. Based on past experience, some ideas are given below as to general issues which should be addressed. These issues are addressed from the applicant's point of view but much of the information is equally applicable to the evidence of an objector with the necessary changes and additions including a critical review of the applicant's evidence.

The applicant's submission can best be divided into three main areas: first, basic information about the applicant; second, the services it proposes to operate; and third, an economic analysis. The applicant should bear in mind three essential questions that will require an answer:

1. Is there a need and demand for the proposed services?
2. Is the applicant the person who should provide those services?
3. Can it do so economically and without causing unacceptable damage to any existing carriers?

Unless it can persuade the CAA through its evidence (both in its written submission and at the hearing) that the answer to all these questions is a definite yes then the applicant is unlikely to win its case and be granted the air transport licences for which it has applied.

The applicant

The first part of the submission should deal with the applicant itself giving basic information about any relevant skills and management experience. An established carrier could refer to its past success and reputation in the market place. A new carrier would mention its plans to obtain finance and organise its corporate structure.

The basic concept behind the application should be to demonstrate a corporate strategy of which the proposed routes are an essential part. Perhaps it is part of the start-up plan to create a hub-and-spoke operation with high frequency services in small aircraft or, maybe, part of a carefully constructed major expansion of an international network and so on. It is important that the proposed services are perceived as part of a carefully planned development strategy. Essentially in this first part of the submission the applicant should have established in outline what it is it wants and why.

After this general introduction the actual proposals could be outlined, for example the routes, whether scheduled or chartered, for carriage of passengers and/or cargo and at what times and frequencies. Then the applicant can proceed to explain what is needed and what it plans to provide.

The operation of the services

The second part of the submission should review more thoroughly exactly how the applicant intends to operate the proposed services and why they are needed. At this or some later stage the particular merits of each route must be addressed. It is often appropriate to discuss each point of destination in turn. Such questions as its geographical situation and strategic relevance; economic and commercial importance; cultural heritage; tourist potential; and transportation infrastructure may be relevant and beneficial. This will depend, of course, on what kind of market the proposed services are targeting, for example business travellers, holiday-makers or cargo exporters.

Another section of the submission must show how the applicant plans to operate the services. Much research and preparation will be required for this. There are certain key areas which are discussed below.

Aircraft

The proposal should envisage a particular aircraft and type; its precise configuration (vital information for calculating load factors for the economic analysis dealt with later); and crew requirements. The applicant should explain why it believes this particular aircraft to be most appropriate, what aircraft it has or how it intends to acquire the aircraft and on what basis. For instance, one ploy of objectors has been to suggest that there are no such aircraft of the type specified by the applicant available for it to purchase or lease and without that particular aircraft the proposed operation could not be viable, and so on. A prudent applicant if it has not already obtained the aircraft in its fleet, will have tested the market and found where, when, and at what cost it can obtain the aircraft. Even better it might have agreed the purchase or lease of the aircraft conditional on the licences being granted.

This information should be followed by information showing that the applicant has considered how it would put the aircraft into operation. First, the seat configuration must reflect the classes of service provided. Is it to be a tightly packed economy configuration with short seat pitch for short-haul holiday charters, or a three-class configuration with generous seat pitches for long-haul scheduled flights? It is surprising, but indicative of what detail can be involved, that it has not been unknown for extensive arguments to develop between parties over seat pitch and configuration. Perhaps more significant is the decision as to the different classes of service and how many seats should be dedicated to each. This has a direct effect on load factors, pricing and revenue. Similarly the crew configuration has to be assessed and their appropriate numbers and qualifications for each sector of passengers. Other matters that will need consideration are catering and cleaning (though not the most material of points, questions have been asked about such matters) and the provision of aircraft maintenance and engineering facilities.

Airports

There are, of course, at least two airports to consider on each route and more if there are stop-offs or multi-sector routes. Certain essential information will be relevant to each airport that is used in operating the route. First and foremost, the number and times of proposed movements must be related to 'slots' available. It is no good producing a brilliant proposal to operate to points where the airports are full to capacity. Obtaining 'slots' can be a far from simple matter and the appropriate authorities have to be approached long before the expected date of commencement of operations as 'slot' committees plan months in advance (normally seasonally in six-month periods). Once advice is sought it may be that the proposed schedule of flight times will need to be adjusted. This could in turn affect the demand for those services so it is important that any potential problems with airports are recognised early on in the planning of the services.

The airport facilities will also have to be investigated. It is worth a paragraph or two on such matters as parking; ground handling; passenger handling; catering and facilities in the airport, including check-in facilities; cargo handling and administration facilities; customs and excise and immigration arrangements and security provisions. All these issues and more will have to be thoroughly investigated and eventually arrangements finalised before services can commence.

Experienced operators objecting to a new applicant on an established route will try to catch out the latter in an attempt to show up its inexperience and generally question its ability to operate the services.

Route Analysis

Finally, it will be necessary to produce a certain amount of statistical information by way of numerical schedules and analyses. This should include a full route analysis showing existing traffic and future market forecasts, indicating source and type of traffic. It is necessary to demonstrate that the demand does or will exist for the proposed services. This can mean on any existing route that present capacity is not sufficient to fulfil demand. Ideally, therefore, the new operator would fulfil that excess demand and at best not impact in any way on existing carriers on the route. It may be, however, that present capacity appears to be sufficient for present demand or, at least, an objector is likely to try and argue this. In the case of new routes it is again difficult to assess existing demand when no services are offered. Normally, to some extent, and maybe to a large extent, an applicant will have to make a case for potential demand or growth on the route for which it will provide the extra capacity required. Of course it is also open to an objector, or alternatively a competing applicant, to maintain that it can, and is best suited to provide that capacity.

Whatever the arguments the CAA will want evidence of the existing position on each route, a reasonable forecast for the future and proposals that would provide those future requirements. The CAA will also want to know the impact of the proposed services on any other competitors, basically any other carriers on the route.

Impact

It is worth reminding ourselves at this stage, of one of the CAA's statutory duties in considering air transport licences. Apart from general considerations of securing the provision of air transport services that satisfy all substantial categories of demand at, amongst other things, the lowest charges consistent with a high standard of safety and significantly an economic return to efficient operators, the CAA must specifically consider the effect on other operators (see Section 4 of the Civil Aviation Act 1982). It must have regard to the effect on existing air transport services provided by British airlines of authorising any new services the applicant proposes to provide under the licence and in any case where those existing services are similar (in terms of route) to the proposed new services or where two or more applicants have applied for licences under which each proposes to provide similar services, the CAA must have particular regard to any benefits which may arise from enabling two or more airlines to provide the services in question. Often summarised as 'impact' the effect on other carriers is, therefore, an important area of consideration. An applicant, whether for a route with existing services or a competing applicant, must be careful how it justifies its share of the market (see Section 68 (2) of the Civil Aviation Act 1982).

Applicants have been known to suggest that by licensing them the very increase in services offered will stimulate new demand. Historically it is possible to show that on some routes demand has followed, or at least grown, to fulfil supply. This turns the normal approach on its head by arguing that it is really a chicken-and-egg situation with demand and supply each able to stimulate the other. It is difficult to prove, being at best only a prophecy, but as a concept, if at all valid, it can be of great help to an applicant applying to be licensed on a new or existing route.

Financial data

Quite apart from the route analysis statistics, the applicant must provide detailed information about its internal finances. The CAA will normally require financial projections on a monthly basis for at least two years. (See also the financial fitness requirements discussed in Chapter 6 p. 48.) This information, essentially a business plan, will need to detail the two sides of the financial equation, namely, costs versus revenue, to produce resulting debit or credit balances. As a very general rule of thumb it has been said that the usual results are a loss in year one, break even in year two, and profit in year three. This may be totally incorrect in particular cases but few applicants on new routes find, for example, that they can produce a profit in year one without having their costs or revenue figures attacked as hopelessly optimistic. This is particularly so in start-up situations which experience shows take time to build up as profitable operations.

Within the financial analysis the applicant should forecast all financial expenditure and income. Particularly relevant will be the direct operating costs (known as the 'DOCs') of each route broken down aircraft by aircraft, flight by flight. Detailed forecasts will need to be prepared with regard to passenger and cargo payloads. Projections as to revenue based on these forecasts using proposed pricing tariffs will also allow an average break-even load factor to be calculated. This is the figure representing a percentage of passengers or cargo (as the case may be) required to be carried (in terms of the number of seats or cargo space filled as a percentage of the total capacity, i.e. seats or space available) on the aircraft on a particular flight for costs to be covered and above which a profit will be made. There is often much debate about load factors: in theory the break-even point could be almost anywhere from 1 per cent to 100 per cent but in practice it is often in the area of 70–80 per cent on an established route.

Again objectors can ask many questions about the statistical information presented by the applicant. By casting doubt on costs as being too low and traffic projections in terms of numbers carried as too optimistic, they can often undermine the credibility of the projected results. It may, for instance, be shown that by marginally increasing certain costs and/or by reducing passenger numbers, or both, the break-even load factor will rise considerably and maybe to a level that cannot be supported by passengers carried. In other words, if the projected passenger load factor is lower than a projected break-even load factor the services will clearly be operated at a loss.

These points are often argued at length and it is vital, therefore, for an applicant (particularly as a newcomer on a route where there are existing carriers who are objectors) to have done its homework thoroughly. In terms of costs, existing carriers will have the advantage of actual experience. Traffic forecasts based on growth and other such assumptions are at best prophetic as the future can never be proved in advance. In that sense the applicant can never prove its case in advance but it can produce a financial analysis which appears justifiable and reasonable to the CAA.

International traffic rights

In the case of international routes, an important but separate part of the hearing must consider the bilateral situation; in other words the international traffic rights on the proposed routes. The opinion of the Department of Transport will be sought and its written advice is normally circulated before the hearing (but at no particular time). The parties can then respond in their written submissions to this or leave their contributions to be considered at the hearing.

As will be explained in more detail when discussing the proceedings at the hearing (see p. 77), bilateral considerations are normally subject to the Official Secrets Act and, therefore, that part of the hearing has to be heard in camera (i.e. in private). The circulation of written material relating to these confidential matters also has to be circulated in a restricted manner and advice should be sought before dealing with such information.

The Department of Transport deals with international traffic rights partly because of its confidential and governmental nature and this is, therefore, one of the few areas of aviation regulation not supervised and administered by the CAA. The CAA does, however, have a statutory duty (Section 68 (1) of the Civil Aviation Act 1982) to consider such matters when deciding whether to grant air transport licences: hence the reason it obtains the advice of the

Department of Transport. The Act provides that it is the duty of the CAA to perform its air transport licensing functions in the manner which it considers is best calculated to ensure that British airlines compete as effectively as possible with other airlines in providing air transport services on international routes. In performing those functions it must have regard to any advice received from the Secretary of State with respect to the likely outcome of negotiations with the government of any other country or territory for the purpose of securing any right required for the operation by a British airline of any air transport service outside the UK. It should also have regard to the need to secure the most effective use of airports within the UK.

As will be seen (see chapter 11 p. 79) international relations are also to be considered by the CAA when giving reasons for its decision in respect of transport licences. In the interests of UK national security the CAA can refrain from furnishing the statement of reasons which it is otherwise obliged to do. Thus, whilst the CAA is not involved in international relations which is clearly dealt with by the appropriate government department, the Department of Transport, it must take into account the international position when carrying out certain aspects of its air transport licensing functions.

Timetables

In its proposal an applicant should detail its flight timetable and pricing tariff. The flight schedule must on the one hand relate to potential 'slot' availability at airports of departure and destination (and stopovers if relevant), and marketing targets, on the other. The area of the market which is to be targeted influences the sort of timetable that is required. For instance, holiday traffic would obviously prefer not to travel too early in the morning or late in the day. Business traffic, however, has quite different priorities especially in terms of short-haul flights that might involve travelling there and back within the day. Business travellers want to depart from a nearby airport and leave early in order to arrive in time for morning meetings and to leave after the end of the working day. Thus commuter traffic will have different demands and priorities.

Fares

Similarly affected will be the pricing structures of the services offered. Not only must they relate to a particular target market but also to other competitors on the route. An objector can argue that fares pitched too high will reduce demand, whilst fares that appear lower than those charged by competitors will attract further criticism. Established carriers may argue that their experience shows that the routes cannot be operated profitably at the proposed rates or may claim that the applicant is trying to undercut competitors with predatory pricing. In the 1980s, fear of a price-cutting war, particularly in the aftermath of the Laker experience, was very apparent. The established airlines fear a newcomer will undercut their rates which in turn will force them to take defensive action by reducing their rates and thus start a price-cutting war.

Established airlines argue that this process is destructive to all parties and ultimately nobody benefits – not even the consumer. The established carriers will eventually force the newcomer out of business as they are able to withstand, if necessary, reducing their prices to an unprofitable level until the newcomer can no longer compete and is forced out of the market. In effect it is a commercial war of attrition. Meanwhile, the public gains only a temporary advantage, the established carrier's profitability is affected and eventually the newcomer is destroyed. This syndrome is still presented as an argument against cheap fares. However, it is becoming increasingly difficult to justify in a climate of increasing deregulation and competitiveness. All considerations of pricing should of course be seen in the context of the tariff restrictions which still exist although they have been modified in recent years.

Class of service

In considering all aspects of the product offered, the applicant must also know exactly what different classes of service it will offer. On long-haul scheduled services it will probably be looking at a first class, business class and tourist class. It will have to decide exactly to the last seat how it will apportion the capacity and configure the aircraft. This also introduces the issue of seat pitch; an important variable not only in terms of overall capacity and, at the end of the day, revenue, but also as a potential marketing inducement.

The pricing structure will have to be similarly considered. Particularly for business and first class the extra services, perks and frills will have to be clarified. Such issues in first class can become highly relevant when perhaps price is not so much a decisive factor.

Marketing

Finally, having selected the market to be targeted and described, and the product to be supplied, the applicant must show that the demand for that traffic exists and how it will market its services. First, it will have produced historical and current data on each route, analysing traffic types and volume and explaining changes and trends such as growth and decline. It will then project its own traffic forecasts with explanatory texts to try and justify the analysis. This is essentially the market research background to route analysis. Having established, however, what it thinks will be demanded and what it can supply the applicant must explain how it will sell or market its products. How for instance are the tickets to be sold and what about advertising and promotion? An old trick of objectors is to try to show that the applicant has failed to allow for, or to allow sufficiently for the costs of advertising (which is sometimes considerable in the case of the larger carriers).

The marketing and selling of the proposed services is a vital part of the production process of providing those services. The applicant should provide an indication of the personnel and facilities, both within the organisation and externally, such as travel agents, which will be involved. How the tickets are to be sold and promoted is a key question. Sometimes applicants are able to supply copies of letters from travel agents, tour operators and such like indicating support for the proposed services and their willingness to market them if licensed. This type of letter may be used as evidence of demand (but without the writers of such letters being present for questioning at the hearing, their value is limited).

On international routes the applicant will need to give serious consideration to the international selling network. Will it use independent general selling agents (GSAs) based in each point of destination or plan to install its own offices and employees? The latter is obviously a far more costly exercise and whatever plans are made they will need to be reflected in the cost of the operation portrayed in the economic analysis. The position of a start-up airline is clearly different from that of a large established airline such as a flag carrier who may already have this infrastructure in place, providing an international network of offices and personnel. The costs in that case will probably be already spread over a variety of different operations and this will give an established airline a considerable cost and experience advantage over, say, a start-up airline. These sort of points can be fiercely defended at hearings.

The above hopefully gives an indication to would-be applicants as to the sort of information that should be considered and, where appropriate, supplied in written submissions and on which there will probably be further discussion at the hearing. This information is described here mainly from the viewpoint of an applicant applying for licences on routes it has probably not used before. Naturally the information would have to be changed or adapted for different standpoints. An objector, for example, would need to examine carefully the facts and figures presented by an applicant with a view to criticising their accuracy where possible. An objector already flying a route on which the applicant proposes to fly will be able to compare the other's proposals with its own actual data. This gives the objector the advantage of arguing historical fact as opposed to the applicant's hypothetical forecasts. If possible, objectors in such circumstances try to discredit the applicant's propositions as incorrect due to a lack of knowledge and experience.

Over the years these hearings have been the subject of hard-fought battles. Commercially they are of vital importance, almost a case of life and death, as the question often at stake is, will a particular business operation to be allowed to operate or not. Its effect on the applicant could be a matter of whether it can start or continue an operation and, on the objector, whether it will have competition that might reduce its future profitability.

Witnesses

The written submissions are the main concern during the period leading up to the hearing. Having been produced, certain matters relating to the hearing must be finalised. First, each party must produce one or more persons as a witness who will speak to the submission or a particular part thereof. The witnesses should be prepared to answer questions from the CAA representatives and be cross-examined by the other parties. They will need to be credible and competent in their particular area. The parties may also wish to consider external witnesses to speak as experts on various aspects of their case. This may involve such people as the party's financial advisor, an airline economist or consultant and people representing trade interests, such as travel agents and tour operators who support the party's case.

Second, each party will need to plan carefully how their case will be presented and by whom. As will be discussed in Chapter 11 on procedure at the hearing itself, they can be represented by whomsoever they may choose, from a member of their organisation to an eminent Queen's Counsel (see Regulation 26 (2) of the Civil Aviation Authority Regulations 1991, SI No. 1672, as amended).

The CAA requests in its regulations (Section 9 (1)) of the *Official Record, Series* 1) that it is notified not less than two working days before the hearing of the name and status of the persons who will represent the parties to be heard. Significantly those provisions also make the point that no person other than those with a right to be heard (as defined in Regulation 25 (1) of the regulations) should assume that they will be heard unless they have been notified in writing by the CAA. Persons not entitled to be heard may, nevertheless, be informed of the time and date of the hearing so that, if they wish, they may attend the hearing as observers.

Third, each party must consider its response to the other party's written submission. The applicant may serve a reply to the objector's submission though the time by that stage will be short (as the time by which the objectors should serve their submissions is only ten working days before the hearing).

Furnishing of information by the CAA

Before the date of the hearing the CAA must (in most cases) serve on the relevant people a copy or summary of any information in its possession which has been provided in connection with the case or which the CAA has reason to believe will be referred to at the hearings (Regulation 22). An obvious example is the advice of the Department of Transport in respect to the international traffic right position (see p. 69). The CAA is not, however, obliged to serve any such information in certain circumstances such as information provided by the Secretary of State and certified by him that it would not be in the public interest for it to be disclosed. There are also restrictions on serving information that has been supplied by certain categories of person which in the opinion of the CAA relates to the commercial or financial affairs of the person who has provided it and where disclosure would disadvantage that person to an unwarranted extent. This is a complicated provision and should be read in full where relevant (see Regulation 22).

Preliminary hearings

Before the hearing the CAA may hold a preliminary meeting to discuss the conduct of the case. Notice of the date, time and place must be given to:

(1) every party to the case,
(2) every person who the CAA proposes to hear in connection with the case, and
(3) any person consulted by the CAA (pursuant to Regulation 21) who has responded in writing.

These persons may then attend in person or by their authorised representative. These meetings are to be conducted by a member or employee of the CAA, on its behalf. (See Regulation 23).

11 Procedure at a Hearing

The regulations concerning procedure at hearings are detailed in Regulation 26 of the Civil Aviation Authority Regulations 1991, (SI 1991 No. 1672). Some key points to bear in mind are the following.

Judiciary

The CAA assumes the judicial role of judging the case in accordance with its powers and duties given under the Civil Aviation Act 1982. The regulations provide that a quorum of the CAA for the purpose of conducting a hearing is normally to be two members of the CAA. There are certain exceptional circumstances when the quorum is reduced to one (Regulation 15 (2)).

In practice, representatives of the CAA are appointed to a panel which conducts the hearing. It consists of two members of the Board (of the CAA), who may be full or part-time members and one of whom is the Chairman, in accordance with the regulations. It is these two persons who are responsible for making the decisions. Sitting with them is an advisor, normally a CAA employee connected with the licensing division who can give expert advice. Finally, a Secretary is also appointed to the panel who is normally from the licensing division. It is the Secretary who attends the panel and whose duties usually include drafting the summary of cases for inclusion in the panel's decision.

The CAA has considerable discretion as to how it conducts the hearing. Regulation 26 (1) merely states that the CAA should conduct the hearing sitting with such of its employees as advisors as it thinks fit.

Representation

Every party to the case may either appear in person or be represented by another person. In practice there is a wide choice. Sometimes, for example, airlines may be represented by one of their own employees, who in a large concern may specialise in air transport licensing within the company. Others may instruct solicitors and they in turn may instruct counsel (i.e. barristers) to represent their clients. Legal representation is by no means essential, although depending on the significance of the hearing or strength of any opponents, a person with some advocacy ability is preferable. This skill may not necessarily mean that the person needs to be a qualified barrister and indeed some solicitors have developed such skills. It is, however, sensible to include someone with experience of air transport licensing and some advocacy experience, whether legally trained or not. Some very competent representatives have been developed in-house by airlines. What has to be borne in mind, of course, is that the outcome of these hearings can be of the utmost importance to the operator. The CAA Official Record, Series 1, Section 9 (headed Hearings and Decisions of the Authority) requests that persons to be heard at the hearing notify the CAA not less than 2 working days before the Hearing of the name and status of the persons who will represent them.

The form of the hearing

The outcome of a hearing may provide the means or otherwise to operate routes and therefore, a successful outcome can be of vital importance commercially. The form of the hearing is relatively informal compared to a court of law but follows quasi-judicial procedures. In terms of evidence each party may produce oral or written evidence (Regulation 26 (2)) and may examine any other party to the case, any person the Authority hears (pursuant to Regulation 25 (2)) and any witness produced by any such party or persons.

The procedure of the hearing is as follows. The applicants (who have essentially caused the case to be heard) present their case, normally by way of an opening submission presented by their representative. They then produce their witnesses in turn, each of whom is examined by their representative and then any other party is given the opportunity to cross-examine the witness after which the applicants can re-examine their witness. This process of opening submissions followed by examination-in-chief, cross-examination and re-examination of each witness is a quasi-judicial procedure and gives some order and form to the hearing.

After the applicant's case has been presented, every other party, normally objectors, presents their cases in the same order. Finally, each party makes in turn a closing speech by way of final submission. Where there are several applicants (as may be the case with new routes, for example, the opening of the London City Airport) an order of appearance has to be agreed or dictated by the CAA.

For those who do not wish to be heard at the hearing and therefore make no appearance (provided they have served an objection or representation pursuant to Regulation 20 as already discussed), a written submission may be made instead but this must be served on the authority not less than three working days before the date fixed for the hearing (see Regulation 26 (3)).

Public access

Every hearing is heard in public (Regulation 26 (4)) unless the CAA decides otherwise in respect to part or whole of a particular case. In practice all hearings are in public with the exception of those parts which deal with a bilateral issue. Such matters often involve confidential agreements between nations and other private information which is subject to the Official Secrets Act. Any part of the hearing dealing with these issues must, therefore, be heard 'in camera', essentially in private with just the CAA and the parties present, who are themselves bound by the Official Secrets Act for these purposes. Also subject to these restrictions are any written submissions produced before or during the hearing by any of the parties, the CAA and the Department of Transport. The latter normally submits an opinion as to the foreign traffic-right position on the relevant route. The CAA will note this advice but only in the context of the particular application when considering all the issues involved. It is, of course, not the role of the CAA to deal with bilateral matters but that of the Department of Transport who negotiates such arrangements with foreign governments.

Time and place

A hearing takes place at the time and date set by the CAA. Apart from notifying the parties, official notification is published in the CAA's publication, *The Official Record*, Series 2, which deals with air transport licensing notices. Series 2, produced regularly every week and available by subscription from the CAA's publication office at Cheltenham. (see p. 60), gives notice of hearings. Hearings normally commence at 10am and all persons to be heard are requested to be in attendance not less than 15 minutes before the hearing.

The CAA also states in section 9 of its *Official Record*, Series 1, containing information relating to air transport licensing that, except in cases of unusual urgency, hearing dates will be fixed and announced not less than two months before the hearing itself. These provisions state that once a date has been announced it will be changed 'only for the most compelling reasons'. It is, therefore, of little use for a party to a hearing to complain about the date and try to have it changed once it has been announced (unless there are special justifiable reasons). Personal inconvenience is certainly not sufficient and parties are best advised to keep in contact with the CAA about dates. The CAA have proved themselves to be not unhelpful or inflexible in the past and will normally try to achieve the date that best suits everyone but they are clearly restricted by time pressures and conflicting interests. The CAA diary of hearings has been very full at various times in the 1980s with delays of up to six months and more due to the volume of applications. At quieter times, however, two to three months is more usual (given that a minimum of two months' notice normally has to be given).

Hearings are held at the CAA's premises which span buildings in Kingsway and the adjacent Kemble Street in central London. Normally hearings will take place in the specially formed conference composite on the eighth floor of CAA House.

Recordings

It is normal for all hearings to be recorded by taping equipment and transcribed into written transcripts. The regulations provide that all proceedings at a hearing should be recorded by a shorthand writer or by some other means (see Regulation 26 (7)). If any person requests a record of the proceedings the CAA is obliged to make a record available for purchase at a reasonable price.

There are two exceptions: first, the CAA is only obliged to provide such a record for up to one year after the date of the publication of its decision in the case, and second, such records of the proceedings conducted otherwise than in public only have to be made available for purchase to parties to the case or other persons heard by the Authority at those proceedings. In other words circulation is limited in the case of private proceedings but any person can request a record of public proceedings. In practice, therefore, a request for a transcript or mechanical recording should be made as soon as possible after the hearing to the licensing section (ERG Division), of the CAA. There is also a procedure (under Regulation 26 (8) (b)) for provision of mechanical recordings or transcripts provided to a person having a right of appeal which provides for the establishment of a 'transcript date'. That Regulation 20 provides that the CAA shall publish the decision date and the transcript date.

The decision

The decision of the CAA is given after it has had time to consider the evidence, normally several weeks after the hearing. In practice the time can vary enormously depending on the Authority's work load and such factors as the length and complexity of the hearing.

The CAA has a statutory duty to give reasons for its decision. (Section 67 (2) Civil Aviation Act 1982). Whenever it makes a decision to grant, refuse to grant, vary, suspend, or revoke an air transport licence it must furnish a statement of its reasons for the decision to certain persons. Those persons are the applicant, holder or former holder of a licence as appropriate, and any other person who, in accordance with the regulations, has been an objector in the case or requested such a statement. This is the general rule but there are exceptions. No statement of reasons need be given where first, no such objection has been entered and no such request made, and second, where the decision is taken in pursuance of, and in the terms requested in, an application for the grant of a licence or an application by the holder of a licence for the variation, suspension or revocation of it. In other words, in a situation where there is no justifiable opposition and it would be unnecessary to justify the decision as it is in accordance with what the relevant parties wanted, there is no need to provide a statement of reasons (see Section 67 (2)).

There is another exception to the rule which concerns information that may cause harm to others, notably national security or sensitive commercial information. In particular, if the CAA has reason to believe that the furnishing of a statement of reasons might be contrary to the interests of national security or might affect adversely the relations of the UK with any other country or territory, it is the duty of the CAA to give notice of the case to the Secretary of State and, if he so directs refrain from furnishing the statement or part of it.

In addition, the CAA can refrain from furnishing a statement or part thereof to certain persons if it considers it necessary to do so for the purpose of withholding from that person certain information. This must be information which in the opinion of the CAA relates to the commercial or financial affairs of another person and cannot be disclosed to certain persons without disadvantage to the other person which, by comparison to the advantage to the public and the person to whom it would otherwise be disclosed, is unwarranted (see Section 67 (3)).

As to the form of its written statement regarding decisions in a case, the CAA has a wide statutory authority to publish in such a manner as it thinks fit, particulars of and its reasons for, any decision taken by it in respect of an air transport licence or an application for such a licence (Section 67 (4)).

The Civil Aviation Authority Regulations provide that the CAA shall in its written notification of the decision (to persons having a right of appeal pursuant to Regulation 27 (1)) specify a date, not being less than three working days after the date on which a copy of the notification was available for collection or despatched (which date should be the decision date for the purposes of making an appeal (see Chapter 12)) (Regulation 26 (8) (a)).

12 Appeals

There is a right of appeal to the Secretary of State against the decision of the CAA in respect of an air transport licence application.

Authority

As usual the statutory authority is found in the Civil Aviation Act 1982. Section 67 (5) states that the Secretary of State has a duty to make regulations giving prescribed persons a right of appeal from any decision of the CAA with respect to or to an application for an air transport licence. The regulations may contain such provisions as the Secretary of State thinks fit. The Secretary is also specifically authorised on an appeal to direct the CAA to reverse or vary the decision in question and to do or refrain from doing such other things as may be specified in the direction. The Secretary is, however, to have regard to certain particular duties of the CAA imposed by Sections 4 and 68 of the Act.

Section 4 states the general objectives of the CAA which form the vital underlying principles by which it is to exercise its powers. These objectives are stated in full in Chapter 6, p. 45. In brief they include protecting the interests of users balanced by ensuring an economic return to efficient operators. Section 68 specifies the duties of the CAA in exercising its air transport licensing functions. These are considered in more detail in Chapter 6.

Regulations

The regulations made pursuant to the statutory requirements are The Civil Aviation Authority Regulations 1991 (S.1 No. 1672) (as amended by statutory instruments from time to time). Regulations 27, 28 and 29 deal with appeals to the Secretary of State and reference should be made to the actual text for specific compliance. The general terms are specified below.

Right of appeal

Normally every party to a case before the CAA has the right of appeal to the Secretary of State from the CAA's decision with respect to an air transport licence or an application for a licence. This means that objectors as well as applicants can appeal. Regulation 27 states two categories of exception as specified in (e) and (f) of Regulation 25 (1) which lists the persons entitled to be heard before a decision is made (as discussed in Chapter 9). In conclusion, therefore, the following parties can appeal:

(a) the applicant;
(b) the holder of any air transport licence;
(c) the holder of any air operator's certificate granted under the Air Navigation Order;
(d) the holder of any aerodrome licence granted under an Air Navigation Order.

Time limits

The notice of appeal must be served within 21 days of the decision date (subject to certain exceptions) as stated in Regulation 27 (4). The Regulation states certain exceptions involving cases of allocating scarce bilateral capacity and a request for transcripts when the procedures and timings may be different.

Any person (other than the appellant) serving a written submission on the Secretary of State must normally do so within 14 days of having been served the notice of appeal. The CAA shall furnish the Secretary of State with any submission it may wish to make within 28 days of receiving the notice of appeal. Within 14 days of the expiry of the 28 days the appellant may serve on the Secretary of State a reply to any of the submissions. It should be noted that there are exceptions to this procedure and the above time limits in specified cases (see Regulations 27 and 28). As to the decision, the Secretary of State is not bound to reach a decision within a particular time limit.

Procedure

Notice of appeal

An appeal to the Secretary of State shall be made by notice in writing, signed by or on behalf of the appellant and must clearly identify the case to which it relates and state the grounds on which the appeal is based and the arguments on which it relies (see Regulation 27 (2)).

Service

The appellant must serve the notice of appeal on the following parties specified in Regulation 27 (3):

(a) the Secretary of State;
(b) the CAA;

(c) each of the parties to the case before the CAA;

(d) each person whom the CAA (pursuant to Regulation 25 (2)) has decided to exercise its discretion to hear in connection with the case, whether that person was heard or not; and

(e) any person consulted by the CAA (pursuant to Regulation 21) in connection with the case.

The procedure is fairly strict and those with the right of appeal must exercise that right by particularised written notice served on specific persons within the appropriate time-scale. All parties have the right to serve written submissions but again within appropriate time limits. The appeal is not conducted by means of a public hearing as in the case of contested licence applications before the CAA but by private consideration of the written evidence by the Secretary of State.

Evidence

All evidence as we have seen must be in writing by way of submission in accordance with the regulations. Appellants in their notice of appeal must state any grounds on which the appeal is based and the argument on which they rely (see Regulation 27 (2)). They can then serve on the Secretary of State a reply to any of the submissions served by the other parties. Copies of that reply must be served on the CAA and the parties who have been served with the notice of appeal pursuant to the regulations (see Regulation 27 (8)). Other parties to the appeal and any person served with a notice of appeal pursuant to Regulation 27 (3) (e) may serve evidence by way of written submission on the Secretary of State giving reasons why the CAA's decision should or should not be upheld (see Regulation 27 (6)). Copies of such submissions must be served on the appellant, and the CAA and other parties served with notice of the appeal in accordance with Regulation 27 (6). Time limits must be strictly observed by reference to the Regulations.

The CAA may also serve evidence by way of a written submission on the Secretary of State to include if it thinks fit any amplification and explanation of the reasons for its decision. Copies of its submission must be served on the persons who have been served with the notice of appeal as specified in Regulation 27 (7).

In conclusion, any parties served with notice of appeal have the right to submit written evidence so long as it is served on each of the other parties and in accordance with the regulations. The Secretary of State may request the appellant or any other party who has made a submission, to amplify any point (see Regulation 27 (10)). The Secretary can also obtain further information from the CAA. Significantly, for the purposes of the appeal, no person may submit to the Secretary of State evidence which was not originally before the CAA when it decided the case (see Regulation 27 (11)). Thus no new evidence may be introduced and the appeal must be decided on the original evidence which will be placed before the Secretary of State.

The decision

The Secretary of State may decide as it thinks fit to uphold, reverse or vary the CAA's decision or direct it to rehear the case (Regulation 29 (1)). There are provisions obliging the Secretary to notify the usual parties (see Regulation 29).

Costs

There are provisions in Regulation 29 relating to costs. Significantly, the Secretary of State does have the power, if it thinks fit, to order the appellant to pay to any other party either a specified sum or the taxed amount of any costs. This provision, in particular, is cited to stop a potential appellant commencing an appeal for inappropriate reasons.

It could, of course, be to the commercial advantage of certain parties to try and cause a competitor to delay commencing services on a competitive route. For instance, where a new applicant is licensed an objector to that application who is already operating on that route may know that a few weeks' or a month's delay could cause a newcomer to lose the benefit of the high season or other important timing benefits of that year. It must be remembered though that normally there is no reason why a successful applicant cannot commence operations or otherwise operate in accordance with the original decisions of the CAA even if that decision is being appealed against by other parties. That original decision would normally stand until otherwise reversed.

Technically, then, routes under a new licence can be operated pending an appeal but in practice an operator may feel it is imprudent to commence services which may possibly have to be stopped when the decision of the Secretary of State is known. This can be something of a dilemma to an operator and experience shows that it is particularly hard on start-up airlines whose whole investment and commercial existence depends on operations being commenced.

Anti-competitive behaviour

Regulation 28 deals with appeals from decisions of the CAA after preliminary hearings of allegations of behaviour damaging to a competitor. It essentially provides that Regulation 27 shall apply but with the specific modifications stated in the regulations.

13 Transfer and Surrender of Licences

It is a cardinal principle of the British air transport licensing system that licences are granted to a particular person or body corporate approved by the licensing authority, the CAA, and are not, therefore, freely transferable. Valuable assets they certainly are, as they form the essential right to trade in terms of being able to operate services commercially, but saleable assets they are not, as many an operator has found to its dismay.

In law, though, changes can be made in, say, the ownership of an operator without, technically, the operator itself changing. For example, in the case of a body corporate (or in other words a UK registered company) the shares could be sold to new shareholders so the ownership of the company is totally different though the operator, i.e. the company itself, remains the same named licence holder. Given such situations regulations have been made to protect the essential principle that new persons cannot become licence holders without the approval of the CAA.

Civil Aviation Act 1982

This Act (Section 67 (1)) specifically empowers the Secretary of State to make provision as to the circumstances in which an air transport licence shall or may be transferred or treated as if granted to a person other than the person to whom it was granted.

The Civil Aviation Authority Regulations 1991 (SI 1991 No. 1672)

Regulation 30 deals with the transfer of licences. Subsection (1) is concerned with two particular circumstances: namely, where the sole holder of an air transport licence dies and where a body corporate is involved in a reconstruction or amalgamation.

In respect of the death of a sole holder (being an individual) of an air transport licence the licence is to be treated as if, from the time of death, it had been granted to his legal personal representative. This means that there is an automatic continuity until the application for transfer to a new holder under Regulation 30 (2) is resolved. The definition of 'legal personal representative' for these purposes is a person constituted as executor, administrator or other representative of a deceased person by proxy, administration, or other instrument.

Where a body corporate undergoes a reconstruction or amalgamation and the whole of the business of the licence holder (being a body corporate) or any part of the business relating to the provision of carriage by air for reward of passengers or cargo, is transferred or sold to another body corporate, the licence shall be treated from that date as if it had been granted to that other body corporate. Under Regulation (30) (2), that person (being a body corporate in this case) may then apply to the CAA for the substitution of its own name in the licence instead of the name of the person by whom the licence was held. In respect of the death of a sole licence holder, an application should be made by the legal personal representative for the transfer of the licence to any person (including himself if this be the case) who is entitled to the beneficial interest in the deceased's estate.

An application (made in the above circumstances) should be served on the CAA within 21 days from the day the applicant first became entitled to make it and should state the grounds on which it is based. The vital provision is that if no such application is made within the stated time the licence shall, at the expiration of that period, cease to be treated as if granted to a person other than the person to whom it was granted: in other words the original licence holder. Technically, the aforesaid application is treated as an application for the variation of a licence and such applications and any subsequent appeals shall be subject to the usual regulations relating thereto.

The CAA is always subject to its statutory obligations and, in particular, as provided in Section 65 (2) and (3) of the Civil Aviation Act, 1982 (relating to the suitability of the applicant).

In conclusion, therefore, in the event of either the death of a sole licence holder or a reconstruction or amalgamation affecting a corporate licence holder, contingency regulations exist to temporarily transfer the licence to allow for applications by the new holder to be made within a specific time. If these applications are not made the temporary transfer is terminated. It is essential, therefore, that immediate action is taken when such events occur or, where possible, consideration is given in advance.

It should always be remembered that the CAA has a continuing duty to ensure that the basic qualifications for a licence holder, specified by the Civil Aviation Act 1982 (such as nationality, financial and other fitness requirements) continue to apply (see Section 65 of the Act). In order to assist it in this function the CAA normally requires that individual operators supply it with updated information on a regular basis (particularly in respect of their financial position).

Surrender of licences

Regulation 31 of the Regulations deals with the surrender of air transport licences by the licensing authority, the Civil Aviation Authority. Where revocation or variation of the licence has taken effect the CAA may require that the licence be surrendered for cancellation or variation as appropriate. Anyone who fails to so comply without reasonable cause is guilty of an offence and liable on summary conviction to a fine. Again the authority of the CAA is enforced by criminal penalties in the event of default or non-compliance.

Continuation of expired licences

Apart from the grant of a new licence or a change to an existing one there is also the situation where an existing licence expires and the holder wishes to renew it for a further period. Section 65 (6) of the Civil Aviation Act 1982 provides that where a holder of a current licence applies for the grant of another licence in continuation of or in substitution for the current licence and does so before the expiry of the term of the current licence then it shall not cease to be in force by reason only of the expiry of that term (so long as the application is made within the time limits laid down by the CAA, as referred to below). The current licence will, however, expire on the occurrence of one of the following events:

(1) when the CAA gives its decision on the application;
(2) if the CAA refuses the application (or to grant the licence on the terms requested) then when the time expires during which an appeal against that decision may be brought and, if brought, when the appeal is determined or abandoned; or
(3) if such an appeal (pursuant to (2) above) is successful, then the date when the licence granted in consequence to the appeal comes into force.

The statutory provisions referred to above provide that the CAA may specify in a notice published in the prescribed manner, regulations relating to the application by a licence holder for grant of a licence in continuance or in substitution of a current licence. In pursuance of these powers the CAA has given notice in Section 5 of its *Official Record*, Series 1 (headed 'Continuation of Expired Licences') of the times of notification. There are three categories depending on the terms of the licence: not more than six months, more than six months but less than twelve months, and more than twelve months. The period prescribed to each category is half the term of the licence; four months and six months, respectively (see Practice Guide, p. 87). This means that where, for example, a holder of a current licence granted for a whole term in excess of twelve months wishes to make such an application, it should do so not later than the prescribed time (as stated in the 1982 Act) which according to the CAA notice is not later than six months before the expiration of that particular licence.

It is very important, therefore, that holders of air transport licences keep a constant check on the expiry times of their licences and ensure that if they wish to continue these services they apply to the CAA for the grant of a new licence within the time specified. Otherwise, they will almost certainly have to cease operating the services specified in that licence by the correct time of its expiry date.

Practice guide

A holder of an air transport licence should note the following:

1. On expiry of the licence, the licence will not be automatically or otherwise renewed. There are no renewal provisions.
2. If the holder wishes to continue those services operated under that licence it must apply to the CAA for a new licence.
3. In order to try and ensure continuance of the services when the existing licence expires, the holder must apply for a new licence within the time limit specified by the CAA.

Deadlines

In order to take advantage of the statutory provisions allowing the licence to continue in force for a limited time, an application should be made by the holder of a licence for a term within one of the three categories (see the left-hand column below) not later than the corresponding period prescribed in the right-hand column.

Term of licence	Period prescribed
Not more than 6 months	1/2 the term of the licence
More than 6 months but not more than 12 months	4 months
More than 12 months	6 months

14 Air Operator's Certificate

Central to any regulatory structure must be the person or body responsible for compliance. This figure in air transport is normally the 'operator' and it is the operator who must show competence in order to obtain an air operator's certificate – the first stage up the regulatory ladder.

Before looking at the regulatory requirements themselves we should consider how the term 'operator' is defined. The definition of 'operator' is found within the Air Navigation Order (SI 1989 No. 2004) as amended, under the interpretation section, Section 106 (3). References in the Order to the operator of an aircraft refer to that person who at the relevant time has the '*management of that aircraft*'. It should be noted that the word 'management' is used, in other words the person responsible for the aircraft and its activities at a particular time as distinct from, say, ownership or other such interests. Those interests may, of course, to some extent be detached from the daily operation of the aircraft and it is clear that the aim of the regulatory requirements is to hold accountable the person or body who manages or, in effect, controls the aircraft at the relevant time.

It should also be noted that the definition of operator in the Civil Aviation Act 1982 (for the purposes of that Act) is essentially the same, namely, 'in relation to an aircraft, means the person having the management of the aircraft for the time being, or in relation to a time, at that time' (see the general interpretation section of the Act, Section 105).

For the purposes of the Air Navigation Order, Article 106 (3) continues to clarify this definition in circumstances where the aircraft is subject to a hire or loan situation. The Order states that for the purposes of the application of any provision in Part III (relating to airworthiness and equipment of aircraft) when, by virtue of any charter or other agreement for the hire or loan of an aircraft, a person (other than an air transport undertaking or an aerial work undertaking) has the management of an aircraft for a period not exceeding 14 days the foregoing provisions of the paragraph shall have effect as if that agreement had not been entered into. In other words, where an aircraft is, say, hired out for a period of fourteen days or less, the original operator remains the operator (for these purposes) and not the new hirer of the aircraft. If the period of the hire, loan or charter is for over 14 days then the operator may well change to the person or body who hires, charters or takes on loan the aircraft so long as they fall within the definition of operator for the relevant times. The significance of these time limits could, therefore, be of considerable inportance. An operator, say, hiring an aircraft out for a short period of under 14 days should be particularly aware of the inherent dangers of that situation and should consider carefully a situation where it may be liable as operator although it may not have been actually operating the aircraft at the relevant time.

Part II of the Air Navigation Order 1989 (SI 1989 No. 2004 amended as of 1 January 1991) headed 'air operator's certificates' provides for the issue of such certificates. Article 6 (1) states the basic requirement for certification and Article 6 (2) deals with the issue of certificates.

Aircraft registration

The requirement for a certificate is related to aircraft registration, thus making it obligatory for aircraft registered in the UK to be subject to an air operator's certificate. Specifically, a UK registered aircraft is prohibited (under Article 6 (1)) of the Order) from flying on any flight for the purpose of public transport otherwise than under and in accordance with the terms of an air operator's certificate granted to the operator of the aircraft. The certificate must certify that the holder is competent to secure that the aircraft operated by it on the relevant flights are operated safely.

This reiterates the fundamental point of certification which is to secure the safety of aircraft operations for the purpose of public transport. It should be noted that the requirements defined refer to 'any' flight for the purpose of 'public transport'; there is no limitation in terms of place, for instance, and a relevant flight could take place in any part of the world. The definition 'any' is wide: its one restriction is that the flight must be for the purpose of public transport. Public transport is defined at some length in Article 107 of the Order (as are the relevant definitions of 'aerial work' and 'private flight'). Article 107 (2) states that an aircraft in flight shall be deemed to fly for the purposes of public transport if a series of considerations are thereto applicable. In general terms, this includes where valuable consideration is given or promised for the carriage of passengers or cargo in the aircraft on a relevant flight. In any particular case reference should be made to Article 107 of the Order which defines in considerable detail the difference between categories of public and private flight.

Issuance

The CAA is the issuing authority and it is to the CAA that applications for an air operator's certificate should be made. Before granting a certificate the CAA must be satisfied that the applicant is competent.

Competency

In assessing an applicant's competence the CAA must have regard, in particular, to the following:

(1) previous conduct and experience;
(2) equipment;
(3) organisation;
(4) staffing; and
(5) maintenance and other arrangements.

The CAA must be satisfied that the prospective operator is competent to secure the safe operation of aircraft of the types specified in the certificate. The CAA has authority to grant certificates subject to any conditions (such as aircraft type, specified flights) or purposes, that it thinks fit.

The certificate will remain in force for the period specified therein subject to the overriding provisions expressed in Article 66 of the Order. Article 66 in general terms gives the CAA the right, if it thinks fit, provisionally to suspend or vary any certificate, licence, approval, permission, exemption, authorisation, or documunent isssued, granted or having effect under the Order, pending enquiry or consideration of the case. Thereafter, the CAA may if it is satisfied that sufficient grounds have been shown after due enquiry, revoke, suspend or vary any such certificate (licence, approval, permission, exemption, authorisation or other document). The holder must then surrender the same to the CAA within a reasonable time of it being required to do so.

Practice guide

Enquiries and applications should be made to the Civil Aviation Authority as the body authorised to issue the certificate.

1. As of 1 January 1991 the initial or general enquiries can be made in writing or by telephone to: Civil Aviation Authority, Safety Regulation Group, Aviation House, South Area, Gatwick Airport, Gatwick, West Sussex, RH6 OYR. Tel: (0293) 567171.

2. In response the prospective applicant will normally receive a standard AOC package including:

   (a) CAA form 1240 – application form for an air operator's certificate.
   (b) CAA document No. 36 headed 'The completion of application forms for the grant or variation of an air operator's certificate'.
   (c) Further information and notes including fees charged for application.

3. The application form should be completed and returned together with the appropriate fee to the address, as notified.

15 Aircraft Registration

The next few chapters will deal with a number of major issues affecting aircraft, commencing with their registration, followed by the regulation of airworthiness and finally addressing some key problem areas affecting aircraft such as accident investigation and various rights of detention in aircraft.

The Chicago Convention of 1944

As we have already seen the Chicago Convention dealt with a number of fundamental issues affecting aircraft in the operation of international civil air transport. In particular, Chapters III and V of the Convention, respectively, deal with nationality and conditions to be fulfilled in respect of aircraft.

Nationality of aircraft

In the context of other issues discussed at the Convention, concerning both the definition of national rights and the imposition of uniform international standards, it was essential to lay down certain basic principles on the nationality, registration and marking of aircraft. The fundamental principle enshrined in the Convention is that aircraft are to have the nationality of the state in which they are registered (Article 17). Logically, they can have only one nationality. Dual registration is not recognised and the Convention specifically states that aircraft cannot be validly registered in more than one state. Registration may be changed from one state to another so long as the aircraft is only registered in one state at any particular time (see Article 18), thereby changing the nationality of the aircraft. The registration or transfer of registration of an aircraft in any contracting state is to be made in accordance with its laws and regulations.

The Convention, therefore, lays down the fundamental principles of nationality upon which further regulations are based. For example, we have seen that the regulations requiring air operator's certificates and air transport licences have based their applicability on the nationality of the aircraft in question, namely, aircraft registered in the UK in respect of these regulations as enforced under this jurisdiction.

Marking

Under the Convention every aircraft that is engaged in international air navigation shall bear its appropriate nationality and registration marks (Article 20). This, of course, is the origin of the current international system of aircraft marking. Each mark consists of a series of letters, normally five, of which the first one or two identify the country and the rest designate the particular aircraft.

Registration

As we have seen from Article 17 in Chapter III of the Convention, a system of registration is assumed in that it forms the basis of nationality. We have also seen that registration may be changed from state to state. Article 21 obliges each contracting state to undertake to supply on demand certain information to another contracting state or the International Civil Aviation Organisation (ICAO). This information concerns the registration and ownership of any particular aircraft registered in that state. In addition, each contracted state is obliged to furnish reports to ICAO in accordance with the regulations. ICAO may so prescribe, 'giving such pertinent data' as can be made available concerning the ownership and control of aircraft registered in that state and habitually engaged in international air navigation. The data obtained by ICAO is to be made available by it on request to other contracting states (Article 21).

Further provisions in relation to aircraft are prescribed in Chapter V of the Convention. These cover such areas as documentation to be carried; radio equipment; certificates of airworthiness; personnel licensing; recognition of certificates and licences; and journey log books. Again the Convention lays down the basic general principles as a foundation on which individual states will base their own regulatory structure.

Having established these general principles laid down by the Chicago Convention, we should now look at the national system of regulations constructed on these international foundations. The main regulations in respect of aircraft registration are found in the Air Navigation Order 1989. Before looking at the regulations themselves we should note the usual progression from the establishment of principles by an international convention to their adoption and enforcement by legislative procedures produced by individual states. Section 60 of the Civil Aviation Act 1982 provides for Orders in Council, thereafter referred to as 'an Air Navigation Order' for carrying out the terms of the Chicago Convention and for regulating air navigation generally. Thus, the authority to give effect to the Chicago Convention in the UK is given by that statute and the means to enforce it is an air navigation order. There have been a series of such orders made and updated as and when necessary, the current order being the Air Navigation Order 1989, as amended.

The 1982 Act also specifies in Section 60 (3) (a) that an air navigation order may contain provision as to the registration of aircraft in the UK.

Air Navigation Order 1989

Part I of the Order, (SI 1989 No. 2004) in Articles 3, 4 and 5, deals specifically with the registration, marking and nationality of aircraft. For precise compliance the text itself should be consulted.

First, Article 3 deals with the requirement to be registered. It states the general rule that an aircraft shall not fly in or over the UK unless it is registered in one of three places specified as:

(1) some part of the Commonwealth; or
(2) a contracting state (i.e. a party to the Convention); or
(3) some other country in relation to which there is in force an agreement between Her Majesty's Government in the United Kingdom and the Government of that country which makes provision for the flight over the United Kingdom of aircraft registered in that country.

There then follows lists of exceptions (for example, kites, captive balloons or special cases involving gliders). Significantly, for instance, an aircraft may fly unregistered on a flight which begins and ends within the UK without passing over any other country, i.e. a purely internal flight, so long as it is in accordance with 'B conditions' as set out in Schedule 2 of the Order. Schedule 2 consists of a detailed list of A and B conditions, the latter found in the schedule section following the last article of the Order, Article 109. To ensure compliance careful reference should be made to the actual text.

It should be noted that contravention of paragraph (1) of Article 3 can be serious. If an aircraft flies over the UK in contravention of the said provisions and in such a manner or circumstances that had the aircraft been registered in the UK an offence under the Order or any regulations made thereunder would have been committed, the like offence shall be deemed to have been committed in respect of that aircraft. This fills a loophole ensuring that those committing offences under the Order cannot escape liability because technically the relevant aircraft was not registered when it should have been in accordance with the Order.

The next article, Article 4, deals with the procedure and technical aspects of the register. Again, a review of the actual text is essential for the purposes of proper compliance. The Article provides in some detail for the actual working of the registration system. Some major aspects are considered below.

(1) The CAA is established as the authority keeping and maintaining the register. This is in fact done from its main premises in Kingsway, London. (Enquiries should be addressed to the CAA at CAA House, 45–59 Kingsway, London WC2B 6TE.)
(2) The information to be kept on the register includes the following:

   (a) the number of the registration certificate;
   (b) the nationality mark of the aircraft and the registration make assigned to it by the CAA;

(c) the name of the constructor of the aircraft and its designation;

(d) the serial number of the aircraft and: (i) the name and address of every person who is entitled as owner to a legal interest in the aircraft or share therein, or, in the case of an aircraft which is the subject of a charter by demise, the name and address of the charterer by demise; and (ii) in the case of an aircraft registered in pursuance of paragraphs (4) or (5) of this article (relating to unqualified residence and charters), an indication that it is so registered.

In short, the register records various identification marks, the manufacturer and particulars of owners or similar interests.

(3) The person in whose name the aircraft is registered is known as the registered owner and shall receive a registration certificate.

(4) Applications for registration should be made in writing to the CAA and should be acompanied by: (i) appropriate evidence relating to the ownership and chartering of the aircraft, and (ii) particulars of the aircraft including a proper description classified in column 4 of the general classification of aircraft in part A of Schedule 1 of the Order (a standard form of application is reproduced in Appendix II, p. 267).

(5) Aircraft only qualify for registration in the UK register if they fall within the national qualifications set out in Article 4 (2) of the Order. This specifically states when an aircraft cannot be registered, or continue to be registered. In very general terms this prohibits registration when:

(a) an aircraft is registered outside the United Kingdom; or

(b) an unqualified person holds any legal or beneficial interest by way of ownership; or

(c) registration in some other part of the Commonwealth is more suitable; or

(d) it would be 'inexpedient in the public interest'.

The last category of exclusions obviously allows wide scope for refusal on the grounds of national interest.

So far the qualifications have been expressed in negative terms (detailing why registration cannot be affected on the United Kingdom register) and by reference to *aircraft* whereas Article 4 (3) then turns to *persons* who may qualify. It is expressly stated that only certain persons 'and no others' should qualify to hold a legal or beneficial interest by way of ownership in an aircraft registered in the UK, or a share therein. Again the list is specific. It includes the following:

(a) The Crown in right of Her Majesty's Government in the UK.

(b) Commonwealth citizens.

(c) Citizens of the Republic of Ireland.

(d) British protected persons.

(e) Bodies incorporated in some part of the Commonwealth and having their principal place of business in any part of the Commonwealth.

(f) Firms carrying on business in Scotland. 'Firm' has the same meaning as in the Partnership Act 1890.

Thus the essential criteria for qualification is nationality.

In order to provide for circumstances where registration may be appropriate but not within the strict definition of Article 4, paragraph (3), paragraphs (4) and (5) deal with two hybrid, but not unusual, practical situations.

Article 4 (4) basically provides that an aircraft may be registered in the United Kingdom where an *unqualified* person holds a legal or beneficial interest by way of ownership in that aircraft (or a share therein) if:

(a) that person resides or has a place of business in the United Kingdom; and

(b) the CAA is satisfied that the aircraft may otherwise be properly so registered.

This is subject to the overriding condition that the aforesaid person shall not cause or permit the aircraft to be used for the purpose of public transport or aerial work. In other words this exemption would normally only apply where the aircraft was used for private purposes. It would, for instance, be useful for foreign nationals residing in the United Kingdom who wish to base and operate from there an aircraft for their own private use: clearly they would wish to have that aircraft registered in the United Kingdom. Likewise a foreign national could be a corporation having a place of business in the United Kingdom wishing to register its private aircraft.

Paragraph (5) deals with charters which naturally can produce some unstraightforward situations. Again, like the preceding paragraph, paragraph (5) allows for registration in situations involving unqualified persons. In general terms it provides that where the aircraft is 'chartered by demise' to a qualified person, the CAA may register the aircraft in the name of the charterer (irrespective of whether or not an unqualified person is entitled as owner to a legal or beneficial interest therein) so long as it is satisfied that the aircraft may otherwise be properly so registered. This registration may remain for the continuance of the charter, subject, of course, to the provisions of the article.

(6) Article (4) also deals with changes in ownership including aircraft transfer. Paragraph (10) makes it clear that if at any time after registration an unqualified person becomes entitled to a legal or beneficial interest by way of ownership in the aircraft or a share therein the registration shall (subject to the special provisions in paragraphs (4) and (5)) become void and the certificate of registration must be returned forthwith by the registered owner to the CAA. The onus is, therefore, on the registered owner to act properly and appropriately in respect of any changes that would affect registration in accordance with the Order.

(7) Accordingly the registered owner has a duty to inform forthwith the CAA in writing of certain changes, namely:

(a) any change in the particulars of the aircraft (which were furnished to the CAA on application for registration);
(b) the destruction, or permanent withdrawal from use of the aircraft;
(c) in the case of an aircraft registered pursuant to paragraph (5), the termination of the demise charter.

There is also an overriding obligation on any person becoming the owner of a registered aircraft to notify the CAA in writing within 28 days.

Additional provisions within Article 4 deal with other matters such as the CAA's right to update or change the register, changes to the regulations by the Secretary of State and registration of aircraft subject to undischarged mortgages.

It should be noted that an interest by way of ownership is not specifically defined but is clarified in paragraph (15) as not to include an interest in an aircraft to which a person is entitled only by virtue of their membership of a flying club. References in paragraph (11) to the registered owner include in the case of deceased persons their legal personal representative and in the case of a body corporate which has been dissolved, its successor. Despite these provisions, paragraph (16) states that nothing in Article 4 shall require the CAA to cancel the registration of an aircraft if, in its opinion, it would be inexpedient in the public interest to do so. Again the CAA has an overriding power to protect the public interest.

Nationality and registration marks

Article 5 of the Order deals with both these matters. Essentially the Order enforces the general principles of the Chicago Convention which require aircraft to take the nationality of the country in which they are registered and to bear appropriate nationality and registration marks (if engaged in international air navigation).

Article 5 prohibits an aircraft (other than one permitted under the Order to fly unregistered) to fly unless it bears painted or fixed thereon, in the manner required by the law of the country in which it is registered, the nationality and registration marks required by that law. Hence all such aircraft should clearly display such marking which enables the onlooker to establish their nationality and registration marks. Aircraft registered in the UK must bear marks specified as appropriate in Part B of Schedule 1 to the Order.

Conversely an aircraft is not to bear any marks which purport to indicate it is registered in a country when it is not so registered or that indicate it is a state aircraft of a country when it is in fact not so (unless that country has sanctioned it). In other words an aircraft should not bear marks which it is not entitled to.

16 Aircraft Mortgages

The mortgaging of aircraft is an essential function in a commercial world. As valuable and expensive assets, aircraft can rarely be acquired without some form of outside financial assistance and any such funding is normally secured by way of a charge on these assets. The financing of aircraft acquisitions is a huge subject in itself particularly in the area of leasing with its elaborate financing arrangements often involving complicated international tax planning. Aircraft finance as such will not, therefore, be addressed in this book. On the regulatory side, however, the reader's attention is addressed to the registering of aircraft mortgages. The Civil Aviation Act 1982 authorises the making of provisions for the mortgaging of aircraft registered in the UK (or capable of being so registered) (Section (86)). The provisions are to be made by Order in Council and are produced by statutory instrument.

The current statutory instrument is the Mortgaging of Aircraft Order 1972 (SI 1972 No. 1268) as subsequently amended. The Order describes the regulations in respect of the registration of aircraft mortgages and should be carefully consulted in any particular case. Some of the essential provisions are indicated below.

Security

The basic rights regarding the use of UK registered aircraft as security are set out in Article 3 of the Order. Any aircraft registered in the UK nationality register may be made security for a loan or other valuable consideration. This can also include such an aircraft with any store of spare parts for that aircraft.

Register

The CAA maintains a register of aircraft mortgages on which may be entered a mortgage of an aircraft registered on the United Kingdom nationality register (Article 4 (1)).

Registration

Applications to register a mortgage should be made to the CAA by or on behalf of the mortgagee in the form set out in the Order (see Appendix II p. 271). The application should be accompanied by the appropriate fee and a copy of the mortgage duly certified by the applicant to be a true copy (Article 4 (2)).

Where two or more aircraft are the subject of one mortgage, or alternatively one aircraft is subject to two or more mortgages, separate applications should be made in respect of each aircraft or of each mortgage, as the case may be (Article 6 (1)).

Mortgages in languages other than English should be translated. A copy of the translation duly certified by the applicants as being to the best of their knowledge and belief a true translation should accompany the copy of the mortgage sent with the application to the CAA. When an application to enter a mortgage or priority notice (see below) is duly made the CAA should enter it on the register by placing the application form therein and noting both the date and time of entry (Article 7 (1)). The latter may be important to establish priority if two or more applications are recieved on the same day. Significantly, applications duly made are entered in the register in order of their receipt by the CAA (Article 7 (2)). It may well be advisable for the sake of speed and certainty that applications are hand-delivered to the CAA in particular circumstances.

Once entered the CAA must notify the applicant of the time and date of entry on the register (of the mortgage or priority notice as the case may be) and of the registered number of the entry. A copy of this notification is to be sent by the CAA to the mortgagor and to the owner (Article 7 (4)).

The opening hours and days of the registry are to be specified by the CAA in its *Official Record*. Any application delivered when the office is closed will be treated as having been received immediately after the office is next opened (Article 7 (3)).

Priority notice

There is a procedure establishing priority in respect of contemplated applications. Known as a priority notice, this is a notice of intention to make an application to enter a contemplated mortgage on the register. Applications to enter a priority notice should be made to the CAA by or on behalf of the prospective mortgagee in the form set out in the Order and accompanied by the appropriate fee (Article 5).

Discharge

The CAA will make an entry on the register to the effect that the mortgage has been discharged on receipt of the appropriate documentation. To achieve this an application in the form provided by the Order duly completed and signed by or on behalf of the mortgagee, together with a copy of the document of discharge or receipt for the mortgage money, or any other document which shows (to the satisfaction of the CAA) that the mortgage has been discharged must be received by the CAA (together with any appropriate fee). The CAA must then enter the form of discharge on the register and mark the relevant entries on the register 'Discharged'. Thereafter, the CAA must notify the mortgagee, mortgagor and the owner that this has been done (Article 9, as amended).

Inspection

Public inspection of the register is obviously vital for the protection of prospective mortgagees and others with rights or potential rights in the assets of the business. Under the Order any person may, on application to the CAA and payment of the appropriate fee, inspect any entry in the register (specified in the application). Inspection can be made in person (see practice guide p. 101) or by sending the application for inspection and the fee to the CAA who on receipt will supply to the applicant a certified copy of the entries on the register specified in the application. The registry is open on the days and hours specified by the CAA in the *Official Record*.

There is also a procedure for seeing whether or not there is an entry on the register with respect to a particular aircraft. A person may wish to find this out before undertaking a full search against an aircraft. The CAA on receiving an application with the appropriate fee, will notify the applicant as to whether or not there are any entries in the register relating to any aircraft specified in the application by reference to its nationality and registration marks.

A document purporting to be a copy of an entry in the register shall be admissible as evidence of entry if it purports to be certified as a true copy by the CAA.

Priority of mortgages

It is obviously essential to establish the order of priority where there is more than one mortgage relating to the same aircraft. This could be important in the case where, for example, two or more mortgagees try to enforce their rights in the security, which is, in fact, one particular aircraft. It would then be necessary to establish who has priority in respect to their various, probably conflicting rights.

The general rule under the Order (see Article 14 (1)) is that the mortgage of an aircraft entered in the register shall have priority over any other mortgage or charge on that aircraft, other than another mortgage entered in the register. This is subject to certain rules applicable to mortgages made before 1 October 1972.

Where two or more mortgages of an aircraft are entered in the register, those mortgages shall as between themselves have priority according to when they were respectively entered in the register. This is subject to allowances for priority notices and specifically, again, for mortgages made prior to 1 October 1972 (Article 14 (2)).

In the case of priority notices, where such has been entered in the register and the contemplated mortgage referred to in the notice has been made and entered in the register within 14 days, that mortgage shall be deemed to have priority from the time when the priority notice was registered. In reckoning the period of 14 days there is to be excluded any day which the CAA has specified in its *Official Record* as a day on which its office is not open for the registration of mortgages (Article 14 (4)). In other words the 14 day period consists of 14 office days (as defined by the CAA).

Nothing in these provisions relating to priority of mortgages is to be construed as giving a registered mortgage any priority over any possessory lien in respect of work done on the aircraft (whether before or after the creation of registration of the mortgage) on the express or implied authority of any persons lawfully entitled to possession of the aircraft or over any right to detain the aircraft under any Act of Parliament (Article 14 (5)). For example, if when a mortgagee tries to enforce a security it is found that the aircraft is already being detained by an aerodrome authority for unpaid airport charges (see Section 88 of the Civil Aviation Act 1982) the rights of that detainor normally take priority over those of the mortgagee.

Indemnity

The Order provides that where a person suffers loss by reason of any error or omission in the register or of any inaccuracy in a copy of an entry in the register supplied pursuant to Article 11 (2) (relating to the supply by the CAA of certified copy of entries on the register) or a notification made pursuant to Article 11 (3) (relating to the inspection by the CAA of the register for entries on a particular aircraft) they shall be indemnified by the CAA. This is subject to two exceptions where this indemnity will not be payable: first, where the person has suffered loss which is caused by, or substantially contributed to by their fraud or has derived title from a person so committing fraud, and second, on account of expenses incurred in taking or defending any legal proceedings without the consent of the CAA (see Article 18).

Miscellaneous

The main provisions have been discussed but there are also a number of additional provisions which relate to other matters. For example, provisions for the rectification of the register by a court order; references to the applicability of other legislation (for example bills of sale and company legislation) and in relation to Scottish regulations and criminal sanctions for the furnishing of false information and forgeries. Significantly, the Order states that all persons shall at all times be taken to have express notice of all facts appearing in the register, but the registration of a mortgage shall not be evidence of its validity (Article 13).

Practice guide

1. The register is kept by the CAA at 45–59 Kingsway, London WC2B 6TE. The current (as of 1 January 1991) opening times are 10am to 4pm, Monday to Friday inclusive but excluding bank holidays, etc. Telephone enquiries can be made on 071 832 6299.
2. Searches can be made in person (at the above premises) or by sending a written application.

17 Aircraft Airworthiness

Fundamental to the safe operation of international civil aviation is the complex structure of regulations relating to aircraft safety. These commence with the basic principles set down in the Chicago Convention of 1944.

The Chicago Convention

Chapter V of the Chicago Convention, headed 'Conditions to be Fulfilled with Respect to Aircraft', lists among these criteria certificates of airworthiness and operating crew licences. Fundamentally, every aircraft engaged in international navigation is to be provided with a certificate of airworthiness issued or rendered valid by the state in which it is registered (Article 31). In other words, UK registered aircraft will be certified by the CAA under the laws of the UK. The Convention provides (Article 33) that certificates of air-worthiness and competency and licences issued or rendered valid by the contracting state in which the aircraft is registered shall be recognised as valid by the other contracting states, provided that the requirements under which they were issued or validated are equal to or above the minimum standards which may be established from time to time pursuant to the Convention.

This provision epitomises the main functions and aims of the Convention. It confronts the problem of conflicting laws and jurisdictions of different states. Whilst recognising the independence of individual states with their own legal jurisdictions and providing for their recognition by others, it tries to ensure overall uniformity, or at least a system of uniform standards, to be established in the future.

In Chapter VI, headed 'International Standards and Recommended Practices', the Convention deals first and most importantly with the adoption of international standards and procedures (Article 37). Under this provision each contracting state actually undertakes to collaborate 'in securing the highest practicable degree of uniformity in regulations, standards, procedures, and organisation in relation to aircraft, personnel, airways and auxiliary services in all matters in which such uniformity will facilitate and improve air navigation'. This is to be achieved through the vehicle of the International Civil Aviation Organisation (ICAO) which adopts from time to time, as necessary, international standards and recommended practices and procedures over a wide range of matters. Specific matters are listed but are not inclusive in that the provision is widely drafted to include 'such other matters concerned with the safety, regularity, and efficiency of air navigation as may from time to time appear appropriate'.

Civil Aviation Act 1982

The next link in the legal chain is, of course, the United Kingdom statute book and we turn again to the Civil Aviation Act 1982. As previously mentioned, Section 60 of the Act gives the power to give effect to the Chicago Convention. Basically it allows the provisions of the Convention to be in force in the United Kingdom jurisdiction by means of an air navigation order made pursuant to the Act. The Act goes on to state that the air navigation order may contain provisions that are 'requisite or expedient' for carrying out the Convention and any annex or amendment thereto.

The Act then lists such matters that the air navigation order may contain. Included in this list are provisions for prohibiting aircraft from flying unless certificates of airworthiness issued or validated under an air navigation order are in force and upon compliance with conditions relating to maintenance or repair (see Section 60 (3) (b) of the Act).

Air Navigation Order 1989

So far, therefore, following the usual legal chain we have seen the source of international regulation in the Chicago Convention made effective in the UK jurisdiction by the Act under which the specific provisions will be made. These provisions currently in force are found in the Air Navigation Order 1989, as amended (SI 1989 No. 2004). Part III deals with airworthiness and equipment of aircraft.

Requirement

Article 7 establishes the requirement for a certificate of airworthiness. In particular an aircraft is prohibited from flying unless there is in force a certificate of airworthiness in respect of it. The certificate must be duly issued or rendered valid under the laws of the country in which the aircraft is registered and any conditions subject to which the certificate was issued or rendered valid must be complied with.

There are, however, some exceptions to this general requirement of certification. The specific exceptions all apply to flights beginning and ending in the UK without passing over any other country and which also fall within one of the following categories:

(a) a glider, if it is not being used for the public transport of passengers or aerial work (other than aerial work which consists of the giving of instruction in flying) or the conducting of flying tests in a glider operated by a flying club of which the person giving the instruction or conducting the test and the person receiving the instruction or undergoing the test are both members;

(b) a balloon (if it is not used for the public transport of passengers);

(c) a kite;

(d) an aircraft flying in accordance with the 'A conditions' or the 'B conditions' set forth in Schedule 2 to the Order; or

(e) an aircraft flying in accordance with the conditions of a permit to fly issued by the CAA in respect of that aircraft.

The 'A conditions' and 'B conditions' refer to two lists of detailed and specific conditions applicable to particular aircraft. These provisions should be carefully studied by a prospective operator and are found in Schedule 2 of the Order.

It should be noted that the basic requirement for airworthiness certification is in respect of any aircraft, and the regulations commence 'an aircraft shall not fly unless there is in force in respect thereof a certificate of airworthiness...'. It is all embracing and not restricted to UK registered aircraft. Such aircraft are addressed at the end of the general requirements. In the case of an aircraft registered in the United Kingdom the certificate of airworthiness is to be a certificate issued or rendered valid in accordance with the provisions (of Article 8) of the Order.

Issue

The CAA is the issuing authority for certificates of airworthiness. Before issuing a certificate of airworthiness in respect of any aircraft it has to satisfy itself that the aircraft is fit to fly having regard to the following:

(1) the design, construction, workmanship and materials of the aircraft (including in particular any engines fitted therein), and of any equipment carried in the aircraft which it considers necessary for the airworthiness of the aircraft; and

(2) the results of flying trials, and such other tests of the aircraft as it may require.

Provided that the CAA has already issued a certificate of airworthiness in respect of an aircraft, which in its opinion is a prototype aircraft (or a modification of a prototype) it may dispense with flying trials in the case of any other aircraft if it is satisfied that it conforms to such prototype or modification (Article 8 (1)).

The certificate of airworthiness must specify the categories of transport

(Article 8 (2)) and purposes for which the aircraft can be flown (Article 8 (2)). It may also specify the performance group to which the aircraft belongs (Article 8 (4)) and may generally be subject to any other conditions the CAA thinks fit (Article 8 (3)).

Foreign certificates

As regards certificates of airworthiness with respect to any aircraft which are issued under the law of any country other than the UK, the CAA may, subject to such conditions as it thinks fit, issue a certificate of validation thus rendering valid (for the purpose of this Order) a certificate of airworthiness issued outside the United Kingdom (Article 8 (5)).

Termination and renewal

A certificate of airworthiness or renewal issued under these provisions shall remain in force for such periods as may be specified therein (Article 8 (6)). The CAA may renew a certificate of airworthiness from time to time for such a period as it thinks fit (see Article 8 (6)). This is always subject to the overall authority of the CAA provisionally to suspend or vary any certificate (including a certificate of airworthiness) licence, approval, permission, exemption, authorisation or other document issued, granted, or having effect under the Order pending enquiry into or consideration of the case and could lead to further suspension or variation or revocation (see Article 66).

A certificate of airworthiness or validation shall cease to be in force with respect to particular aircraft in the following circumstances:

(1) if the aircraft (or such of its equipment as is necessary for the airworthiness of the aircraft) is overhauled or repaired or modified, or if any part of the aircraft or of such equipment is removed or replaced, otherwise than in a manner and with material of a type approved by the CAA either generally or in relation to a class of aircraft or to the particular aircraft;

(2) until the completion of any inspection of the aircraft or of any such equipment as aforesaid, being an inspection made for the purpose of ascertaining whether the aircraft remains airworthy and (i) classified as mandatory by the CAA; (ii) required by a maintenance schedule approved by the CAA in relation to that aircraft; or

(3) until the completion to the satisfaction of the CAA of any modification of the aircraft or of any such equipment as aforesaid, being a modification required by the CAA for the purpose of ensuring that the aircraft remains airworthy.

These are the basic requirements as to certification ensuring that an aircraft cannot fly for most purposes unless it is airworthy to a minimum international standard.

Aircraft, as machines, by their very nature need constant maintenance and repair. To issue a certificate of airworthiness one day to a suitably qualifying aircraft has a diminishing effect in that it does not give any assurance that that aircraft will continue to be maintained at that standard. It is essential, therefore, that in order to maintain standards there are provisions for continuing maintenance. The basic regulations in the United Kingdom are to be found in Article 9 which states that an aircraft registered in the UK in respect of which a certificate of airworthiness in either the transport or the aerial work category is in force shall not fly unless:

(1) the aircraft (including in particuar its engines) together with its equipment and radio station, is maintained in accordance with the maintenance schedule approved by the CAA in relation to that aircraft; and
(2) there is in force a certificate (in the Order referred to as a 'certificate of maintenance review') duly issued in respect of the aircraft which certifies the date on which the maintenance review was carried out and the date when the next review is due.

The maintenance referred to in (1) above must specify the occasions on which a review must be carried out for the purpose of issuing a certificate of maintenance review (Article 9 (2)). The certificates will only be issued by qualified persons being those specified in the Order (Article 9 (3)). It is obviously essential that only suitably qualified and licensed personnel are responsible for aircraft maintenance and the issue of certificates of maintenance review. Such a person is not to issue a certificate of maintenance review unless they have first verified that:

(1) maintenance has been carried out on the aircraft in accordance with the maintenance schedule approved for that aircraft;
(2) inspections and modifications required by the CAA as provided in Article 8 of the Order have been completed as certificated in the relevant certificate of release to service issued in accordance with Article 11 of the Order;
(3) defects entered in a technical log of the aircraft in accordance with Article 10 of the Order have been rectified or the rectification thereof has been deferred in accordance with the procedures approved by the CAA; and
(4) certificates of release to service have been issued in accordance with Article 11 of this Order;

and for this purpose the operator of the aircraft shall make available to that person such information as is necessary.

Finally, the certificate itself is to be issued in duplicate and one copy (of the current certificate) must be carried in the aircraft (as required in Article 61) and the other must be kept elsewhere by the operator (Article 9 (5)). These certificates should normally be kept preserved by the operator of the aircraft for a period of at least two years after issue (Article 9 (6)).

These regulations form the backbone of the continuing maintenance requirements for aircraft registered in the UK. They essentially give effect to the intentions behind the general principles stated in Chicago in 1944. The Order, of course, does little more than reflect these principles and as any aircraft operator or engineer knows they are the tip of the iceberg. For beneath these simple requirements lies a plethora of detailed technical specifications and requirements which support the overall objective, that of safe aircraft maintained to a high international standard. These detailed regulations and technical requirements obviously need to be consulted by any person intending to operate or look after aircraft.

It is not appropriate for a book of this kind to delve into the world of technical manuals and the readers' attention is drawn to the main principles so they can establish a framework of general understanding. Many of these regulatory matters concerning an airline operator originate from the Chicago Convention and it is from that essential document that one can trace so many of the main principles of aviation.

18 Aircraft Documentation

Before leaving this area of aircraft regulation, one should perhaps refer to another principle laid down by the Chicago Convention. It concerns the essential documents relating to an aircraft and certain regulations in relation to them. Article 29 of the Convention states that certain documents must be carried in every aircraft of a contracting state engaged in international navigation. The documents specified are listed below:

(1) certificate of registration;
(2) certificate of airworthiness;
(3) appropriate licences for each member of the crew;
(4) journey log book;
(5) if the aircraft is equipped with radio apparatus, the aircraft radio station licence;
(6) if it carries passengers, a list of their names and places of embarkation and destination;
(7) if it carries cargo, a manifest and detailed declarations of the cargo.

There is a particular duty (Article 34) to maintain a journey log book in respect of every aircraft engaged in international navigation. In that log book must be entered particulars of the aircraft, its crew, and of each journey, in such form as may be prescribed from time to time pursuant to the Convention.

Documents to be carried

The Air Navigation Order 1989 (as amended) pursuant to its powers under the Civil Aviation Act 1982, takes up the principles of the Convention, and thus provides the basic regulations applicable in the United Kingdom with effect to aircraft-related documentation.

Part VII of the Order deals solely with 'documents and records'. Its first provision (Article 61) provides that an aircraft shall not fly unless it carries the documents which it is required to carry under the law of the country in which it is registered. This reflects the main principle stated by the Convention, though it should be noted that it is not identical and, in fact, takes the principle somewhat further. The Convention states a mandatory duty to carry the specified documents but this is limited to aircraft of a contracting state, i.e. parties to the Convention who are, of course, the only persons who can be bound in what is essentially a contractual relationship. The UK Order is more far-reaching in that it, in theory, extends to all flying aircraft, though in that context it restricts its application not to specifically-named documents (as does the Convention) but to those required to be carried under the law of the country in which the aircraft is registered. This is a different angle from that of the Convention. It attempts to provide more extensive control, but in so doing has to rely on the requirements of individual states.

Clearly the UK authorities have no jurisdiction over other states. They would presumably, however, be able to prohibit a foreign-registered aircraft from flying within the UK jurisdiction if it failed to comply with the documentary requirements.

The second regulation laid down by the Order deals specifically with aircraft registered in the UK and it makes additional requirements on them. Every aircraft registered in the UK must, when in flight, carry documents specified in the Order (under Schedule 11). This applies unless the flight is intended to begin and end at the same aerodrome and does not include passage over the territory of any country other than the UK, in which case the documents may be kept at that aerodrome (instead of being carried in the aircraft) (Article 61 (2)).

Apart from this exception, Schedule 11 specifies some ten categories of documents to be carried in an aircraft. Which of these is to apply depends on the type of flight operated, being one of four categories. Specific reference should be made to the detailed requirements to ensure the appropriate documents are carried. As a very general indication the documents likely to be required on a flight for the purpose of public transport are as follows:

(1) a radio licence pursuant to the Wireless Telegraphy Act 1949 and tele-communications log book;
(2) certificate of airworthiness;
(3) flight crew licences;
(4) load sheet;
(5) certificates of maintenance review;
(6) technical log;
(7) operator's manual;

and if the flight is defined as international air navigation, the following are also required:

(8) certificate of registration; and
(9) copy of the pilot's notified procedures in relation to intercepted aircraft.

Records

Having dealt in principle with the documents to be carried on an aircraft, the Order deals with records that must be kept. It provides (Article 62) that the operator of public transport aircraft registered in the UK shall, in respect of any flight by that aircraft which may fly at an altitude of more than 49,000 feet, keep a record (in a manner prescribed) of the total dose of cosmic radiation to which the aircraft is exposed during the flight, together with all the names of members of the crew during that flight.

Production of documents

The Order then deals with the production of documents and records and specifies when particular persons are obligated to produce certain documents. Such documents can only be requested by authorised persons (see Article 63). An authorised person also has the power to inspect and copy any certificate, licence, log book, document or record which that person has the power, pursuant to the Order and any regulations made thereunder, to require to be produced (see Article 64).

Preservation of documents

The Order also provides regulations for the preservation of certain documents. It includes provisions for continuity to ensure that should an operator of an aircraft cease to be an operator for whatever reason (including death) the documents should continue to be preserved by the appropriate people as specified in the Order (see Article 65).

Non-compliance

Finally, there are provisions in respect of the non-compliance with certain regulations relating to documentation. For example certain deceptions, intentional damage or false entries are forbidden (Article 68). Perhaps not so obvious is the provision that all entries made in writing in any log book or specified records must be made in ink or indelible pencil (see Article 68).

19 Crew Licensing

Crew licensing is an essential part of maintaining safety and other standards in the operation of aircraft. The basic principles are, as usual, laid down by the Chicago Convention of 1944.

The Chicago Convention

Article 32 of the Convention provides that the pilot and all other members of the operating crew of *every* aircraft engaged in international navigation should be provided with certificates of competency and licences issued or rendered valid by the state in which the aircraft is registered. Again the provision is related to where the aircraft is registered and it should be noted that here the requirement relates to every aircraft engaged in international navigation and is not expressly restricted (as in many provisions in the Convention) to contracting states, i.e. parties to the Convention. Considering, however, that the Convention is essentially a contractual document between states, its provisions clearly do not bind in any way states or persons who are not a party to it. It essentially promotes a system of international standards that would hopefully be adopted by as many states as possible.

Article 32 also states that each contracting state reserves the right to refuse to recognise for the purposes of flight above its own territory, certificates of competency and licences granted to any of its nationals by another contracting state.

Air Navigation Order 1989

The Air Navigation Order 1989 (as amended) sets out the basic regulation of crew licensing in the UK. Part IV deals with 'Aircraft, Crew and Licensing' in particular. The first essential principle is that an aircraft shall not fly unless it carries a flight crew of the number and description required by the law of the country in which it is registered (Article 19). Obviously this could be difficult, or indeed impossible, to enforce in areas where the UK has no jurisdiction. The second principle (Article 19 (2)), therefore, specifically regulates aircraft registered in the UK, which must carry a flight crew adequate in number and description to ensure the safety of the aircraft and of at least the number and description specified in the certificate of airworthiness issued or rendered valid under the Order, or, if no certificate of airworthiness is required under the Order, the certificate of airworthiness (if any) last in force under the Order in respect of that aircraft.

Pilots

A key area of regulation is the number of flying crew. For instance, for most commercial operations the minimum number of pilots is two. More specifically, the Order states that:

> A flying machine registered in the United Kingdom and flying for the purpose of public transport having a maximum total weight authorised exceeding 5,700kg. shall not carry less than two pilots as members of the flight crew thereof.

In respect of aircraft having a maximum total weight authorised at 5,700kg or less there is a relatively new ruling. On and after 1 January 1990 such an aircraft, registered in the UK and flying for the purpose of public transport in circumstances where the aircraft command is required to comply with the Instrument Flight Rules and powered by certain categories of engine, should also carry not less than two pilots as members of the flight crew. The applicable engines are one or more turbine jets or others, as specified (Article 19 (3)).

Navigators

With regard to flight navigators, an aircraft registered in the UK engaged on a flight for the purpose of public transport must carry (subject to the exceptions below) either a flight navigator as a member of the flight crew or navigational equipment approved by the CAA and used in accordance with any conditions subject to which that approval has been given. These requirements are only applicable on routes where the aircraft is intended to be more than 500 nautical miles from the point of take-off measured along the route to be flown, and to pass over an area specified in Schedule 7 of the Order (see Article 19 (4)). Schedule 7 gives detailed lists by way of points of longitude and latitude.

Radio operator

A flight radio operator must also be carried in an aircraft registered in the UK in circumstances where it is required to be equipped with radio communication apparatus (in accordance with the Order) (see Article 19 (5) and Article 14).

These are the basic requirements for the provision of pilots, navigators and radio operators. In addition, the CAA is specifically empowered, if it appears to be expedient to do so in the interests of safety, to direct an operator of a UK registered aircraft to carry additional flight crew (Article 19 (6)).

Cabin crew

Similar provisions are made as to the basic number of cabin crew. Certain flights must include cabin attendants who are to be carried for the purpose of performing in the interests of passengers' safety, duties assigned by the operator or commander of the aircraft and who shall not also act as members of the flight crew (Article 19 (7)).

The flights to which this applies are those which are for the purpose of public transport by an aircraft registered in the United Kingdom and on which is carried 20 or more passengers or which may, in accordance with the certificate of airworthiness, carry more than 35 passengers and on which at least one passenger is carried. In terms of numbers there are to be not less than one cabin attendant for every 50 (or fraction of 50) seats installed in the aircraft, provided, however, that this number can be reduced in accordance with written permission from the CAA. As with the minimum numbers of flight crew expressed in the Order, the CAA may (if it appears to be expedient to do so in the interests of public safety) direct that a particular operator of any aircraft registered in the UK must carry additional cabin attendants as specified (Article 19 (8)).

So far, therefore, we have two established principles within the Order. First, there must be a flight crew on all aircraft registered in the UK and also cabin crew where the flight is for the purpose of public transport with a minimum number of passengers and seats. Second, there must be a minimum number of specified flight crew and cabin crew.

Licensing of flight crew

The Order then proceeds (Article 20) to regulate the licensing of flight crew, an essential part of aviation organisation and safety. The basic requirement is that a person shall not act as a member of the flight crew of an aircraft registered in the United Kingdom unless they are the holder of an appropriate licence granted or rendered valid under the Order. There are a number of exceptions listed (see Article 20) whereby a person within the UK, the Channel Islands and the Isle of Man may act as a flight radio telephone operator, or as a pilot of an aircraft or balloon without being the holder of an appropriate licence. The list covered makes special provision for particular circumstances such as, for example, training or obtaining a new instrument rating.

Subject to those provisions, in accordance with the Order, a person shall not act as a member of a flight crew of an aircraft registered in a country other than the United Kingdom unless the following circumstances apply. Either the aircraft is flying for the purposes of public transport or aerial work and the person holds an appropriate licence (granted or rendered valid) under the law of the country in which the aircraft is registered or, in the case of any other aircraft, the person holds an appropriate licence (granted or rendered valid) under the law of the country in which the aircraft is registered (or under the Order) and the CAA does not in that particular case give a direction to the contrary (Article 20 (2)). It should be noted that with regard to these provisions, separate reference is made to balloons and gliders which have specific requirements.

Grant and renewal of flight crew licences

This subject is dealt with in detail in the Order (Article 21). In very general terms the following points should be noted. The CAA is the authority responsible for the licensing and it has full discretion to grant the relevant licences subject to such conditions as it thinks fit. It must be satisfied that the applicant is a 'fit person to hold the licence' and 'is qualified by reason of his knowledge, experience, competence, skill, physical and mental fitness to act in the capacity to which the licence relates'. For that purpose the applicant is to furnish such evidence and undergo such examinations and tests (including in particular medical examinations) and undertake such course of training as the CAA may require. There are minimum age requirements for each class of licence. A licence shall not be valid unless signed by the holder in ink.

Term

A licence should be granted for the period stated in the licence (subject to the CAA's overall rights of suspension and revocation under Article 66). The CAA may renew the licence from time to time subject to the holder being a fit person and qualified as referred to above. The Order details the requirements, effect and categories of licence and reference should be made in particular to Article 21 of the Order.

Validation

The Order allows validation of non-United Kingdom licences. The CAA may issue a certificate of validation rendering valid for the purposes of the Order any licence of a member of the flight crew of an aircraft granted under the law of any country other than the UK. Such a certificate may be issued subject to whatever conditions and for what period the CAA thinks fit (Article 22).

Log books

The Order provides that every member of the flight crew of an aircraft registered within the UK, and every person who engages in flying for the purpose of qualifying for the grant or renewal of a licence under the Order should keep a personal flying log book. This log book must record the following information:

(1) the name and address of the holder;

(2) particulars of the holder's licence (if any) to act as a member of the flight crew of an aircraft; and

(3) the name and address of the holder's employer (if any).

The holder must record particulars of each flight including such details as:

(1) the date and places at which the holder embarked and disembarked (on and from the aircraft) and the time spent during the course of the flight when the holder was acting in his capacity as a member of the flight crew (or for the purpose of qualifying for the grant renewal of a licence);

(2) the type and registration marks of the aircraft;

(3) the capacity in which the holder acted in flight;

(4) particulars of any special conditions under which the flight was conducted, including night flying and instrument flying; and

(5) particulars of any test or examination undertaken while in flight.

It should be noted that particulars should be recorded of any test or examination undertaken in a flight simulator. For the provisions relating to log books see in particular Article 23 of the Order.

Flying instruction

The Order provides essentially that a person cannot give any instruction in flying (for the purposes of obtaining a pilot's licence or any inclusion or variation of the rating in a licence) unless they themselves are properly licensed and qualified. Article 24 gives full details of this provision and its limited exceptions.

The next section of the Air Navigation Order deals with the operation of an aircraft and contains many details of important regulations covering many aspects of aircraft operation from operations and technical manuals to operating conditions, responsibilities of the commander and aerodrome operating minima. Clearly any prospective operator and those whose jobs are involved in aircraft operations would need to familiarise themselves, in detail, with the Air Navigation Order and all other regulations emanating from the regulatory framework set by the Order, pursuant to the 1982 Act.

20 Accidents

Accidents to aircraft, especially those involving the general public and their property, are a matter of great concern to the aviation community. The system of international regulation exists to improve safety and minimise, as far as possible, the risk of accidents but when they do occur there is a web of systems and procedures to investigate and respond to them. As usual, we can trace the general line of regulation from an international source in the Chicago Convention to statutory recognition in the UK under the Civil Aviation Act 1982, and specific regulations made thereunder. In summary the background to the investigation of accidents can be traced as follows.

The Chicago Convention 1944

Article 26 of the Convention lays down the basic principle for the investigation of an accident. Where there has been an accident to an aircraft of a contracting state which occurs in the territory of another contracting state and which involves death or serious injury or indicates serious technical defect in the aircraft or air navigation facilities, the state in which the accident occurs must institute an inquiry into the circumstances of the accident. That inquiry will be in accordance, in so far as its law permits, with the procedure which may be recommended from time to time by the International Civil Aviation Organisation (ICAO).

These are very general provisions but they state two essential principles: first, in certain circumstances there must be an investigation, and second, who is to be responsible for undertaking that investigation. The latter is an important point to establish otherwise there could be at least two states claiming jurisdiction on the inquiry. The Convention also provides that the state where the aircraft is registered is to be given the opportunity to appoint observers to be present at the inquiry and the state holding the inquiry must communicate the report and findings in the matter to that other state.

It is worth noting that the Convention (Article 25) also makes provision for assisting aircraft in distress. Each contracting state undertakes to provide such measures of assistance to aircraft in distress in its territory as it may find practicable and to permit (subject to control by its own authorities) the owner of the aircraft or authorities of the state in which the aircraft is registered, to provide such measures of assistance as may be necessitated by circumstances. Significantly, the undertaking can only be given by contracting states but the duty to provide assistance is not limited to aircraft registered in another contracting state, but presumably any aircraft in distress in the territory of the contracting state.

Finally, the Convention envisages further regulations (normally to be produced under the auspices of ICAO). In this case the Convention provides that each contracting state, when undertaking a search for missing aircraft, will collaborate in co-ordinated measures which may be recommended from time to time pursuant to the Convention.

Since 1944 further international regulations relating to safety and investigation of accidents have been made, both pursuant to Chicago Convention and, in particular, through the vehicle of the ICAO which has, for example, set up an accident and reporting system. By requiring the reporting of certain accidents and incidents it is building up an information service for the benefit of member states.

The European Economic Community (EEC)

The EEC has also had an input into accidents and investigations. In particular, a directive was issued in December 1980 encouraging the uniformity of standards within the EEC by means of joint co-operation of accident investigation. The sharing of and assisting with technical facilities and information was considered an important means of achieving these goals. It has since been proposed that a European accident investigation committee should be set up by the EEC (Council Directive 80/1266 of 1 December 1980).

The United Kingdom

In the UK we must look first at the Civil Aviation Act 1982. The Secretary of State is empowered to make regulations in two essential respects. First, the investigation of any accident arising out of, or, in the course of air navigation, occurring in or over the UK or occurring elsewhere to an aircraft registered in the UK; and second, for carrying out any Annex to the Chicago Convention relating to the investigation of accidents involving aircraft, as it has effect from time to time, with any amendment made in accordance with the Convention. The section elaborates further on detailed matters which may be included in such regulations.

Accident is defined by the Act as including any fortuitous or unexpected event by which the safety of an aircraft or any person is threatened (Section 75 (4)). Section 75 makes provision for regulations relating to a wide variety of accident situations. For example, notification of accidents, prohibiting interference to aircraft, and the cancellation, suspension, endorsement, of surrender of any licence or certificate granted under an Air Navigation Order or Section 62 of the Act (relating to provisions for the control of aviation in war and emergencies).

The regulations made pursuant to the Act in respect of civil aircraft are found in the Civil Aviation (Investigation of Air Accidents) Regulations 1989, (S.I. 1989 No. 2062) as amended from time to time. These regulations are reasonably detailed and should be reviewed in full by persons involved in the investigation of an accident. For our purposes, attention is drawn to a selection of points.

First, an accident is defined as including an incident and a reportable accident. A reportable accident means an occurrence associated with the operation of an aircraft which takes place between the time when any person boards the aircraft with the intention of flight, and such time as all such persons have disembarked. During that time any of the following incidents will be construed as a reportable accident: where

(1) any person suffers death or serious injury while in or upon the aircraft, or by direct contact with any part of the aircraft (including any part which has become detached from the aircraft) or by direct exposure to jet blast, except when the death or serious injury is from natural causes, is self-inflicted, or is inflicted by other persons or when the death or serious injury is suffered by a stowaway hiding outside the areas normally available in flight to the passengers and members of the crew of the aircraft; or

(2) the aircraft incurs damage or structural failure, (other than either engine failure or damage, when damage is limited to the engine its covering or accessories, or damage limited to propellers, wing-tips, antennae, tyres, brakes, fairings, small dents or puncture holes in the aircraft skin), which adversely affects its structural strength, performance or flight characteristics and which would normally require major repair or replacement of the defective components; or

(3) the aircraft is missing or is completely inaccessible.

'Serious injury' is defined as an injury which is sustained by a person in a reportable accident and which either:

(1) requires a stay in hospital for more than 48 hours within 7 days from the date on which the injury was received; or

(2) results in a fracture of any bone (except simple fractures of fingers, toes, or nose); or

(3) involves lacerations which cause nerve, muscle, tendon damage, or severe haemorrhage; or

(4) involves injury to any internal organ; or

(5) involves verified exposure to infectious substances or injurious radiation;

(6) involves second or third degree burns or any burns affecting more than five per cent of the body surface.

These definitions are obviously important to operators and commanders of aircraft who must ensure that the regulations are enforced. The regulations make provision for the reporting of reportable accidents (Article 5) and it is the duty of the commander of an aircraft, or if he is killed the operator, to give full details, as specified, of the accident. Information relating to an accident may be published by the Chief Inspector (Article 6).

The purpose of accident investigation is stated in Article 4 of the regulations.

> The fundamental purpose of investigating accidents under these regulations shall be to determine the circumstances and causes of the accident with a view to the preservation of life and the avoidance of accidents in the future; it is not the purpose to apportion blame or liability.

When a reportable accident occurs in or over the United Kingdom, access to the aircraft is normally restricted to authorised persons (otherwise than for specified reasons such as removal of animals, particular contents or certain safety measures) (Article 7). Most importantly, the regulations provide for the Secretary of State to appoint persons as inspectors of air accidents, one of whom should be the Chief Inspector of Air Accidents (Article 8 (1)). It is the Chief Inspector who determines whether or not an investigation into an accident should be carried out (Article 8 (2)). The powers of the inspectors are various (see Article 9). In particular, they have considerable powers of access and examination over relevant aircraft, extended to certain rights to enter and inspect other places and buildings. Inspectors may take statements from all such persons as they think fit and significantly have power, by summons under their hands, to call before them and examine all such persons as they think fit and require those persons to answer any question, furnish any information, or even produce any document. Generally an inspector is given the power 'to take such measures for the preservation of evidence as he considers appropriate'. Inspectors, therefore, have considerable powers to enable them to investigate an accident to an aircraft pursuant to the regulations. Indeed, they could be seen as aircraft-accident policemen for these purposes.

All investigations are to be in private (Regulation 10). Upon completion of the formal investigation, the Chief Inspector should submit to the Secretary of State the report of the inspector who carried out the investigation (Regulation 11). Upon completion of a field investigation the Chief Inspector shall submit to the CAA such information as it considers desirable in the interest of the avoidance of accidents in the future.

Reports submitted to the Secretary of State (upon completion of a formal investigation) must state the facts relating to the accident followed by analysis of the facts and conclusions as to the causes of the accident together with any recommendations which the inspector thinks fit to make with a view to the preservation of life and the avoidance of accidents in the future. Before such reports are submitted the inspector must, where it appears practicable to do so, serve notices on the operator and commander of the aircraft and any other person whose reputation is, in the inspector's opinion, likely to be adversely affected by the report. These persons have a right to make representations in reply which must be in writing and served on the inspector within 28 days of service of the notice (Article 12 (3)). The inspector must then consider such representations before the report is finally submitted to the Secretary of State.

Any person served with the aforesaid notice must be served by the Chief Inspector with a copy of the report (Article 12 (4)). It is prohibited to disclose or permit to be disclosed any information contained in a notice or report without the prior consent in writing of the Chief Inspector (Article 12 (5)).

Part II of the regulations deals in some detail with review boards. In short, the review board is a procedure by which persons served with an inspector's notice (referred to above) may be heard in their own defence where they feel they might be adversely affected by the report. Thus, any persons served with a notice may within 21 days serve on the Secretary of State a written notice (referred to as a Notice of Review) that they wish those findings and conclusions in the report from which it appears their reputation is to be adversely affected, to be reviewed by the review board. The regulations provide for the setting up of such a board and lay down the procedures by which it will consider the evidence of the people before it. Afterwards, unless there are good reasons to the contrary, the Secretary of State shall cause the inspector's report and the report of the review board to be made public as it thinks fit.

Part III of the regulations deals with the holding of public enquiries. The Secretary of State has the power, where it appears that it is expedient in the public interest to do so, to hold a public inquiry into the circumstances and causes of an accident to which the regulations apply, or into any particular matter relating to the avoidance of accidents in the future. The inquiry is to be held in a quasi-judicial manner under a commissioner who must be either a judge or barrister of not less than ten years' standing and who will be assisted by lay assessors with relevant skills. This 'judiciary' is to be approved by the Lord Chancellor. The rules for the proceedings are laid down in the regulations together with the powers of the commissioner and assessors. The proceedings in many ways follow normal court procedure.

Part IV of the regulations deals with various general matters including provisions relating to accidents to aircraft registered outside the United Kingdom, and provisions for Scotland and Northern Ireland.

These regulations are by no means the only regulations with which an operator and commander should be concerned, in respect of incidents and accidents. For example, these regulations relate only to civil aircraft. There are other regulations, for example the Air Navigation (Investigation of Air Accidents involving Civil and Military Aircraft or Installations) Regulations 1986 (and any subsequent amendments thereof), that may relate to certain accidents, in particular those involving a military aircraft or a collision between a military and civil aircraft. It should also be noted that additional reporting obligations exist in the current Air Navigation Order of 'reportable occurrences' as therein defined (see in particular Regulation 94 of the Air Navigation Order 1989, SI No. 2004).

21 Third Party Rights Against Aircraft

In this section attention is drawn to several common problems which affect the 'owners' or operators' rights in aircraft. We will consider some particular instances where the appropriate authority has the right to detain and even dispose of an aircraft and thus where the rights of the owners and/or operator are overridden. These situations mainly involve the non-payment of airport or navigational charges, Eurocontrol charges or the carrying of contraband.

In all cases the relevant authorities have statutory powers against the relevant aircraft and whilst the measures of detention and even disposal that they can adopt are only effective against contravention or breach of particular obligations, they can have dramatic and serious consequences for those with interests in the aircraft. Unfortunately, the official measures can also affect parties who may not have been the cause of the contravention or breach that led to the action being taken by the authorities. This can result in innocent victims being adversely affected. A number of examples will become apparent when reviewing the nature of these enforcement procedures.

One should always bear in mind the position of the owner or lessor of an aircraft operated by another party who infringes regulations that result in action being taken against that aircraft over which the owner or lessor has basically no control. These situations have to be watched with care by all parties who could, or whose interest could possibly be affected. The first three situations are particularly relevant dangers to lessors and arise out of the non-payment of aerodrome, navigation or Eurocontrol charges. The first of these to be considered are aerodrome charges.

Aerodrome charges

Aerodrome charges are charges rendered by an airport authority for the services it provides at the relevant aerodrome. The Civil Aviation Act 1982 normally refers to aerodrome rather than airport, but each may be taken as synonymous in general usage. The interpretation section of the Act (Section 105) defines aerodrome as meaning:

> any area of land or water designed, equipped, set apart or commonly used for affording facilities for the landing and departure of aircraft and includes any area

or space, whether on the ground, on the roof of a building or elsewhere, which is designed, equipped or set apart for affording facilities for the landing and departure of aircraft capable of descending or climbing vertically.

The airport authority is normally the authority having control and management of the aerodrome. In the UK it is likely to be the (now independently owned) British Airport Authority (BAA) or a local authority. Examples of such charges are parking fees, landing fees, passenger and cargo taxes. They basically arise out of services provided at the aerodrome at ground level as opposed to in-flight navigational services, which are discussed later.

Section 88 of the Civil Aviation Act 1982 gives the relevant authority the right to detain an aircraft for the non-payment of the charges in the following circumstances:

(1) where charges incurred relate to a particular AIRCRAFT (irrespective of the owner or operator at the time), or
(2) where the OPERATOR of the aircraft has incurred charges that relate to other aircraft that he operates.

Section 88 applies to any aerodrome owned or managed by a government department or a local authority (other than a district council in Scotland) and to any other aerodrome so designed by the Secretary of State.

An aerodrome authority is defined as meaning in relation to any aerodrome, the person owning or managing it (Section 88 (10)). Airport charges are defined as charges payable from an aerodrome authority for the use of, or for services provided at, an aerodrome but do not include charges payable by virtue of regulations under Section 73 of the Act. That section provides in some detail for charges for air navigation services and will be considered in more detail later (see p. 124). Section 88 provides that if payment is not made within 56 days of the date when detention begins, the aerodrome authority may sell the aircraft in order to satisfy the charges.

There are a number of safeguards to protect those with interests in the aircraft. First, an aerodrome authority cannot detain or continue to detain an aircraft for alleged default in payment of aircraft charges if the operator or any person claiming an interest therein disputes that the charges, or any of them, are due, or, if the aircraft is detained under Section 88 (1) (*a*) (i) (i.e. charges incurred in respect of that particular aircraft irrespective of operator) disputes the charges were incurred in respect of that aircraft; and gives to the aerodrome authority, pending determination of the dispute, sufficient security for the payment of the alleged charges. Therefore, not only must there be some genuine dispute as to the alleged charges but also an ability to provide adequate security in order to try and prevent detention or continued detention.

Despite the right of the aerodrome authority to sell the aircraft after 56 days as stated in Section 88 (1), there are, therefore, certain protective provisions for those with interests in the aircraft. Apart from the preventative or delaying provisions (under Section 88 (2)), Section 88 (3) provides that the aerodrome authority shall not sell such an aircraft without the leave of the court. Furthermore, the court cannot give leave without proof of the following:

(1) that the sum is due to the aerodrome authority for airport charges;
(2) that default has been made in payment thereof; and
(3) that the aircraft is liable to sale under the section by reason of the default.

This means that the judiciary as an independent authority ensures the objective carrying out of these rather stringent penalty provisions. Further provisions, in Section 88 (4) onwards, lays down a rigid and fair procedure. For example, the aerodrome authority has a duty to notify persons whose interests may be affected by the action of the proposed application and these persons must be given the opportunity of becoming a party to the proceedings. Importantly, if the aerodrome authority obtains the leave of the court to sell the aircraft it must ensure it is sold for the best price that can reasonably be obtained.

Failure to comply with any of the Section 88 (4) requirements in respect of a sale shall not, however, after the sale has been made, be a ground for impugning its validity. In other words, the purchaser of an aircraft in such a sale is protected from attempts to invalidate that sale. Any person, however, suffering loss as a consequence of non-compliance with Section 88 (4) may still have an independent right of action against the aerodrome authority. Section 88 (6) states, in order of priority, the order of application of the proceeds of sale as follows:

(1) in payment of any duty (whether Customs or Excise) chargeable on imported goods or value added tax due in consequence of the aircraft having been brought into the United Kingdom;
(2) in payment of expenses incurred by the aerodrome authority in detaining, keeping and selling the aircraft (including expenses in connection with the application to court);
(3) in payment of airport charges which the court found to be due;
(4) in payment of any charge in respect of the aircraft which is due by virtue of Section 73 of the Act.

(Section 73 relates to charges for air navigation services and is considered below.) Any surplus from the sale once the above have been satisfied is to be paid to the person or persons whose interests in the aircraft have been divested by reason of the sale.

Section 88 (7) extends the power of detention and sale to include equipment of the aircraft and any stores for use in connection with its operation (which are carried in the aircraft). This is irrespective of whether or not they are the property of the person who is the operator of the aircraft in question.

Section 88 (8) deals with any aircraft documents carried in the aircraft which are also subject to the power of detention and which may be transferred to the purchaser if the aircraft is sold. The aircraft documents are defined in Section 88 (10) as any certificate of registration, maintenance or airworthiness of that aircraft, any log book, equipment, or any similar documentation relating to the particular aircraft in question. Court is defined in Section 88 (10) as the High Court in respect of England and Wales and the Court of Sessions of Scotland.

Section 88 (9) clarifies the time and place of detention. The power to detain may be exercised on the occasion on which charges have been incurred or on any subsequent occasion provided the aircraft is on the aerodrome on which the charges were incurred or on any other aerodrome owned or managed by the aerodrome authority concerned.

In conclusion, therefore, Section 88 affords a designated aerodrome authority fairly extensive rights of retention and resale against specific aircraft. The powers have been utilised and should always be borne in mind by, not only operators, but also those with ownership interests in the aircraft. For though such powers are normally exercised only in justifiable circumstances in accordance with the Act, in other words, for unpaid charges, totally innocent parties can be adversely affected. Furthermore, whilst the default continues normally on the part of an operator, an independent owner or lessor who has leased the aircraft to the operator is left in a difficult, almost no-win, situation. The fact that the operator may be in default can jeopardise the property of an innocent owner.

Air navigation service charges

A similar situation can arise for non-payment of charges for air navigation services. The CAA has a clear statutory duty under the Civil Aviation Act 1982 (Section 72) to provide air navigation services over specified areas to the extent that it appears to the CAA that such services are necessary and are not already being provided (by it or others). This gives the CAA fairly extensive powers over air navigation. The specified area covered is the UK and any area outside it for which the UK has in pursuance of international arrangements undertaken to provide air navigational services.

It is also the duty of the CAA to join with the Secretary of State to comply with directions made by the Secretary of State in regard to navigational services. Section 73 of the Act empowers the Secretary of State to make regulations in respect charges for air navigation services provided by the relevant authority at a particular airport.

The current regulations (in force as of 1 April 1991) are the Civil Aviation (Navigation Services Charges) Regulations 1991, as amended from time to time (SI 1991 No. 470). The system and rates of payment are laid down within the regulations. The basic duty to pay for the navigation services provided is specified in Regulation 4 which states that the operator of every aircraft for which navigation services are provided by the CAA in connection with the use of an aerodrome shall pay for those services as detailed within the regulations. The operator is defined here (as in the Air Navigation Order) as the person who at the relevant time has the management of the aircraft.

Regulation 11 deals with detention and sale provisions for unpaid charges. In default of such payments, the CAA or an authorised person may, in accordance with the regulations, take such steps as are necessary to detain, pending payment, either:

(1) the *aircraft* in respect of which the charges were incurred (whether or not they were incurred by the person who was the operator of the aircraft at the time the detention begins), or
(2) any other aircraft of which the person in default is the *operator* at the time when the detention begins.

If the charges are not paid within 56 days of the date when the detention begins, the CAA may sell the aircraft in order to satisfy the charges. These provisions in general, therefore, mirror those under Section 88 of the Civil Aviation Act 1982 in respect of unpaid aerodrome charges. The power to detain and sell for default in payment is also similar. In both cases action can be taken against either the actual aircraft on which the charges are unpaid or another aircraft of an operator who is in default. In each case innocent parties can become involved; for instance, purchasers of aircraft in respect of previous charges, or lessors in respect of lessees' debts and so on.

It is essential, therefore, that any person with an ownership interest in an aircraft over which they do not have operational control should take the appropriate professional advice to consider and, where possible, limit the risks of third party action against their interests. The most obvious example, that of a lessor who has leased out an aircraft to a lessee operator, demonstrates the need for aircraft leases to be drawn up by specialist legal advisers so that all due precautions can be considered and, where possible, taken.

The regulations, having established in Regulation 11 the right of detention and sale, continue from Regulation 12 onwards with provisions as to the procedures, rights and duties of the parties in the event of detention or sale. These are very similar to those in Section 88 of the Civil Aviation Act 1982, in respect of aerodrome charges. In summary, they include the following provisions:

(1) An aircraft cannot be detained or continue to be so detained if the operator of the aircraft or any other person claiming an interest therein disputes any of the charges or, if the aircraft is detained under Regulation 11 (a), (i.e. the actual aircraft in respect of which charges were incurred) disputes the charges relating to that aircraft; and provides sufficient security to the CAA.
(2) The CAA cannot sell the aircraft without the leave of the court which cannot give leave without proof that: a) the sum is due to the CAA for charges under the regulations; b) that default has been made in the payment thereof; and c) that the aircraft which the CAA seeks leave to sell is so liable to sale under the regulations.
(3) The CAA (Regulation 14) must before applying to the court for leave to sell an aircraft take steps to bring the proposed application to the notice

of interested parties (as specified by the Regulations, and in particular by Schedule 2).

It should be noted that Schedule 2 specifies the steps to be taken by the authority to bring to the notice of interested persons the proposed application to court. Further information is given as to the content of such a notice. For instance, in the normal course of events a person in whose name the aircraft is registered should be notified and, therefore, given the opportunity to become a party to the action and take whatever steps they can to protect their interests in the aircraft.

The CAA has a duty to ensure that if leave is given to sell an aircraft that it is sold for the best price that can reasonably be obtained. Failure to comply with any of the requirements of the regulations, while actionable against the CAA, shall not, after the sale has taken place, be a ground for impugning its validity. In other words, the sale cannot be set aside for some irregularity under these regulations, and a third party purchaser is, therefore, protected. However, this does not stop a person who has suffered loss, caused by the CAA failing to comply with any requirement of the regulations, taking action against the CAA itself. If successful in such an action, the harmed party could obviously not obtain return of the aircraft once sold but may be awarded civil damages.

(4) The proceeds of sale of an aircraft are to be applied as prescribed in Regulation 15. In summary, that is, for the following purposes and in this order:

(a) in payment of any customs duty due as a consequence of the aircraft being brought into the United Kingdom;

(b) in payment of expenses incurred by the CAA in detaining, keeping and selling the aircraft which shall include the expenses of the application to court, in other words legal fees and court fees;

(c) in payment of the charges in respect of any aircraft which the court found due from the operator by virtue of these or any other regulations under Section 73 of the Civil Aviation Act 1982; and

(d) in payment of any airport charges incurred in respect of the aircraft which are due from the operator of the aircraft to the person owning or managing the aerodrome at which the aircraft was detained under these regulations.

Any surplus shall be paid to or among the persons whose interests have been divested by reason of the sale.

(5) The powers of sale and detention in respect of an aircraft extend to include the equipment of that aircraft and any stores used in connection with its operations so long as they are equipment and stores carried in the aircraft. This is irrespective of whether or not the property is that of the person who is the operator of the aircraft. So again the rights of an innocent party can be affected.

(6) The power of detention also extends to any of the documents carried in

the detained aircraft. On the sale of the aircraft under these regulations any such documents may be transferred by the CAA to the purchaser.

(7) The power of detention may be exercised on any occasion when the aircraft is on any relevant aerodrome (being one referred to in the table in Regulation 2 of these regulations) or to which Section 88 of the Civil Aviation Act 1982 applies. As at the 1 April 1991 the table referred to lists the following airports in column 1 (with the applicable charges listed in columns 2 and 3):

London – Heathrow; London – Gatwick; London – Stansted; Aberdeen (Dyce); Edinburgh; Glasgow and Prestwick.

Section 88 of the Civil Aviation Act 1982 relates to unpaid airport charges as defined therein (see p. 121).

(8) Nothing in the regulations prejudices any right of the CAA to recover any charges by action. In other words, despite its powers of detention and sale the CAA may use other remedies such as commencing a civil action for debt against the offending party.

22 Detention and Sale of Aircraft for Unpaid Eurocontrol Charges

In addition to the rights we have seen concerning unpaid aerodrome and navigational charges, there are also rights of detention and sale of aircraft by the appropriate authorities for unpaid Eurocontrol charges. The Civil Aviation (Route Charges for Navigation Services) Regulations 1989 as amended from time to time make the appropriate provisions within the UK. (For a general review of Eurocontrol see Chapter 1).

Regulations 10 onwards deal with the special rights of detention and sale where default is made in the payment of charges incurred in respect of any aircraft under the regulations. The organisation (defined in the regulations as the European Organisation for Safety of Air Navigation known as Eurocontrol) may require the CAA to act on its behalf in certain respects. In practice the CAA is in effect the organisation's collecting and enforcement agent in the UK. The regulations state that where such a requirement has been made, the CAA (or authorised person) may, subject to the regulations, take on behalf of the organisation such steps as are necessary to detain, pending payment, either:

(1) The *aircraft* in respect of which the charges were incurred (whether or not they were incurred by the person who is the operator of the aircraft at the time when the detention begins); or
(2) any other aircraft of which the person in default is the *operator* at the time when the detention begins.

If the charges are not paid within 56 days of the date when the detention begins, the CAA may sell the aircraft on behalf of the organisation in order to satisfy the charges.

These powerful provisions are restricted in certain circumstances affording some protection to those whose interests could otherwise be harmed. The CAA (or authorised person) must not detain or continue to detain an aircraft on behalf of the organisation by reason of any default if the operator of the aircraft or any other person claiming an interest therein either disputes that part or all of the charges are due, or if the aircraft is detained under Regulation 11 (A) (i.e. the actual aircraft in respect of which charges were incurred), disputes the charges in question were incurred in respect of that aircraft, and also gives the organisation (pending determination of the dispute) sufficient security for payment of the charges which are alleged to be due.

As with similar rights of sale for unpaid aerodrome and navigational charges the CAA cannot sell the aircraft pursuant to these regulations without the leave of the court. In turn, the court cannot give leave except on proof that the sum is due to the organisation for charges under the regulations, that default has been made in payment thereof, and that the aircraft which the CAA seeks leave to sell on behalf of the organisation is liable to sale by virtue of default under the regulations (Regulation 13). These provisions, therefore, go some way to ensuring the CAA is not cast in the role of both judge and jury and that there must be an independent review by the court subject to objective criteria.

Further protection for interested parties is provided under the notification provision. Essentially the CAA is obliged, before applying to the court for leave to sell the aircraft, to take such steps necessary for bringing the proposed application to the notice of interested persons and affording them the opportunity of becoming a party to the proceedings (see Regulation 14 and Schedule 4). It should be noted that Schedule 4 to the regulations contains extensive provisions as to notification including a list of persons who should be notified when and how. This is, of course, similar to the notification provisions relating to similar proceedings for unpaid aerodrome and navigational service charges.

If the court grants leave to sell the aircraft for the organisation, the CAA has a duty to achieve the best price that can be reasonably obtained. Failure to comply with this or other regulations in respect of the sale, while actionable against the CAA by any person suffering loss in consequence thereof, will not after the sale has taken place be a ground for impugning its validity. This provision protects the purchaser of the aircraft who obviously does not want to risk the sale being overturned because of breaches between the CAA and third parties in these circumstances.

The proceeds of sale are to be applied by the CAA in the following order of priority:

(1) in payment of any customs duty which is due in consequence of the aircraft having been brought into the United Kingdom;
(2) in payment of expenses incurred by the CAA in detaining, keeping and selling the aircraft (including those in relation to the court application);
(3) in payment of charges in respect of any aircraft which the court has found to be due from the operator by virtue of these or any other regulations under Section 73 of the Civil Aviation Act 1982 (this, as we have already seen in Chapter 21, also relates to air navigation charges);
(4) in payment of any airport charges incurred in respect of the aircraft which are due from the operator of the aircraft to any person owning or managing the aerodrome at which the aircraft was detained under these regulations.

Once the sale proceeds have been applied to the above, any surplus must be paid to or among persons 'whose interests in the aircraft have been divested by reason of the sale' (Regulation 15).

The powers of detention and sale are expressed to include equipment of the aircraft and any stores for use in connection with its operation (being equipment and stores carried in the aircraft) whether or not they are the property of the person who is the operator (Regulation 16). The power of detention also extends to any documents carried in the relevant aircraft. Furthermore, such documents may if the aircraft is sold pursuant to these regulations, be transferred by the CAA to the purchaser (Regulation 17).

The power to detain aircraft is obviously restricted to geographical and jurisdictional considerations. The regulations actually state that the power may be exercised on an occasion when the aircraft is at an aerodrome to which Section 88 of the Civil Aviation Act applies. Section 88 (10) applies to any aerodrome owned or managed by any government department, a local authority (other than a district council in Scotland) and to any other aerodromes so designated by the Secretary of State (Regulation 18).

Finally, nothing in these regulations shall prejudice any right of Eurocontrol to recover charges by action. In other words, an offender may be sued for debt rather than action taken against its aircraft. Experience shows that this is sometimes the preferred course and, indeed, in the past Eurocontrol has gone as far as to commence proceedings for the winding-up of a corporate operator for alleged default in payment of its charges.

23 Customs and Excise

The provision of customs and excise facilities at airports to carry out the collection of duties and taxes and enforce certain importation regulations, bans and procedures is another essential piece in the complex airline industry jigsaw.

The Chicago Convention

In brief, provision for customs regulations to address the basic requirements for organisers of civil aviation is provided in the Chicago Convention of 1944. For example, Article 10 provides that a contracting state may require every aircraft which enters or leaves its territory to land at and depart from an airport designated by that state for the purposes of customs and other examination. So apart from a case where aircraft are permitted to cross the territory of a contracting state without landing, that state is in effect able to enforce applicability of its own customs regulations and procedures at its airports. This principle is clarified by Article 13 which states that laws and regulations of a contracting state relating to the admission to or departure from its territory of passengers, crew or cargo of aircraft shall be complied with by or on behalf of such passengers, crew or cargo on entrance, departure or while within the territory of that state.

Examples of these laws and regulations are actually given 'such as regulations relating to entry, clearance, immigration, passports, customs and quarantine'. Customs regulations are, therefore, clearly envisaged as part of the essential regulations and are to be obeyed by the incoming and outgoing aircraft. Interestingly, Article 16 gives specific mention of a contracting state's right (without unreasonable delay) to search an aircraft of another state on landing or departure and to inspect the certificates and other documents prescribed by the Convention. The references to customs regulations are found in Chapter IV of the Convention headed 'Measures to facilitate air navigation'.

In Article 23 each contracting state undertakes, so far as it may find it practicable, to establish customs and immigration procedures affecting international air navigation in accordance with practices established or recommended at the time pursuant to the Convention. This is obviously an attempt, where possible, to unify such procedures on an international basis.

Any further 'practices' envisaged as being introduced pursuant to the Convention are those involving the International Civil Aviation Organisation (ICAO) which, as we have seen, was established by the Chicago Convention for such purposes as a continuing and developing authority. Article 23 also makes it clear that there is nothing to stop the establishment of customs-free airports.

Article 24 makes further provision on customs duty. In particular, certain property which belongs to the aircraft and is retained on board and leaves the territory without being unloaded is to be free from duty: fuel, lubricating oils, spare parts, regular equipment and aircraft stores on board the aircraft in a contracting state. Some parts and equipment imported into a contracting state for incorporation in or use on an aircraft of another contracting state engaged in international air navigation are to be admitted free of customs duty (subject to compliance with the regulations of the state concerned).

Again, as is typical of the Chicago Convention, practical issues are addressed to promote a uniform and organised working system of regulations. The facilitation of getting parts to an aircraft needing maintenance or repair, for example, is an essential function of a working aviation environment. Further reference is made to customs (and immigration) procedures as part of a list of itemised matters in Article 37 for which ICAO is to make and amend from time to time the necessary international standards, recommended practices and procedures. The extensive work of ICAO since 1944 has included the regulation of airports and procedures thereat including customs and excise.

Article 22 makes general provision for facilities to be provided at airports and specifically mentions the administration of laws relating to customs and clearance. Significantly, contracting states are obliged to adopt all practical measures to facilitate and expedite navigation by aircraft between territories of contracting states to prevent unnecessary delays to aircraft, crews, passengers and cargo. This duty to facilitate and expedite to prevent unnecessary delay is perhaps not well known amongst airlines and the travelling public and its potential has certainly not been fully explored.

The United Kingdom

In the UK, detailed provisions are laid down in customs and excise legislation to which specific reference should be made for individual cases and circumstances.

There are general references to customs and excise facilities and regulations within the general aviation framework as we have seen in the Chicago Convention. Reference is also made in the Civil Aviation Act 1982. In particular, the general matters listed in Section 60 which forms the authoritive source of so much of aviation regulation, include customs and excise.

In essence Section 60 (1) and (2) gives power to give effect to the Chicago Convention to regulate air navigation in general. Section 60 (3) provides that an air navigation order (made pursuant to those above powers) may contain certain provisions on a number of matters which are thereafter listed and cover an extensive variety of aviation issues from aircraft to aerodromes, and safety to noise regulations. Paragraph (*m*) covers customs and excise, giving the power to provide within an air navigation order for the time being in force, regulations for customs and excise relating to aerodromes and aircraft, and to persons and property carried therein. Also included are provisions for preventing smuggling by air and provisions, as appropriate, for the protection of the revenue and the importation of goods into the UK without payment of duty.

The main powers of search and retention which can arise in respect of aircraft are found in the Customs and Excise Management Act 1979. Section 141 of the Act gives the Customs and Excise the right, where anything has become liable to forfeiture to seize

> ... any ship, aircraft, vehicle, animal, container (including any article of passengers' baggage) or other thing whatsoever which has been used for the carriage, handling, deposit or concealment of the thing so liable to forfeiture either at a time when it was so liable or for the purposes of the commission of the offence for which it later became so liable.

Traditionally the powers of the Customs and Excise authorities are wide and extensive. Historically this goes back to the powers given them to tackle the extensive problem of smuggling in previous centuries. Today the Customs and Excise has its powers prescribed by statute. Under the said Act it can detain a container including, of course, an aircraft. There is normally one month in which to appeal against the seizure. If the owner does not register an appeal within the given time the property may be forfeit and the Customs and Excise may then be able to sell it and keep the proceeds of sale. There are provisions to mitigate the severity of these powers. For instance, the Commissioners can, if they see fit, deliver up the seized goods to their owner on payment of a sum of money assessed by the Commissioners.

It should also be noted that in using its power of search and retention the Customs and Excise can dismantle a container and is not obliged to put together any such property that it has taken apart. This can, of course, have serious consequences in respect of complicated property such as machinery. For instance, the English customs authorities have the right to search, detain and take to pieces a car returning from a family touring holiday in France in order to investigate the presence of contraband. If the car is then returned to the family the authorities are not obliged to put it together again in the condition in which it was originally produced. This analogy can be continued to an aircraft, though it could clearly reach ludicrous proportions. In reality most property is rarely returned completely dismantled but the potential possibilities are disturbing for property owners.

Clearly an owner or lessor of an aircraft can easily be caught out as an innocent victim in such circumstances, for example, where an owner hires an aircraft to another person who may then use it for the carriage of contraband.

There is very little an innocent owner can do to ensure the speedy return of the aircraft. Indeed, the customs authorities may look directly to the owner for payment of a penalty fee before returning the aircraft. If the owner then reacts, as most owners in that situation do, and refuses, at least initially, to pay as a point of principle there is a risk of the property being permanently alienated from the owner with no guarantee of compensation.

These problems can also relate directly to operators who may be quite innocent though contraband is found on the aircraft they operate and control. On a normal commercial flight many hundreds of people may be involved as consignors of cargo or as passengers with the opportunity to carry in their goods or on themselves, illegal contraband. There is no better example of the practical problems than that of the recent case of *Customs and Excise* v. *Air Canada* [1990] CA. Air Canada operated an aircraft which landed at London Heathrow Airport as part of its usual scheduled service carrying passengers and cargo on a return flight to Canada. At the airport, a container of cannabis resin (a prohibited drug) was found in the discharged cargo by customs officials. In accordance with their powers under the Customs and Excise Management Act 1979, commissioners seized the aircraft on another occasion when it landed at Heathrow on a return flight to Toronto. Considering the same aircraft was due to fly back to Canada with awaiting passengers and cargo within a few hours, the airline was put in a difficult and compromised position. From the point of view of the airline, it had not caused the cannabis to be carried and indeed had no knowledge of its existence in the aircraft and, no doubt, could not see why it should be penalised.

Such legislation, however, is construed as to provide strict liability (a concept discussed in Part III) so that intention, awareness or similar circumstances are irrelevant. Despite this the judge at first instance did hold that to justify seizure under Section 141 (1) of the Act, the plaintiffs had to prove that the defendants knew or ought to have known that the cannabis resin was being carried on that aircraft. Following this judgment, the plaintiffs appealed to the Court of Appeal which in June 1990 reversed the original decision. The Court of Appeal argued that the rights of detention given to the customs authority were exercisable against property (known in law as rights *in rem*) as opposed to rights against a person (*in personam*). The seizure effected by the statutory provisions operated against the aircraft independently of the knowledge, motive or attitude of other persons associated with it. The court further argued that the wording of the section was clear as it stood and did not require that intention on the part of the carrier was in any way present. The condition of the carrier was in fact irrelevant: all that was relevant was the indisputable fact that prohibited drugs were carried in the aircraft, and that the customs authorities had the right to search and detain the aircraft.

In practice, therefore, a carrier can be considerably prejudiced to the extent of potentially serious financial loss caused by events beyond its control. Customs authorities normally act responsibly but this recent case is a reminder of the actual and very real powers they possess and it poses many questions as to their potential application. For instance, what about the passengers, probably regrettably significant in numbers, who carry with them on commercial flights small amounts of drugs as part of their personal property. This could be contained in their checked luggage in the cargo hold or on themselves or in their hand luggage which accompanies them on the aircraft. Questions must be asked as to how far can or would the customs authorities go in attempting to search and detain an aircraft. It can clearly be a problem to airlines who are normally completely innocent victims of such action.

In the Air Canada case, the operator was left with an aircraft legally detained by customs authorities with a load of cargo and passengers awaiting the return flight to Canada. What can an operator do in such circumstances? In many instances the aircraft will be released on payment of fines but these can be considerable and innocent airlines are reluctant to pay them though the commercial pressures on them to do so are considerable. It is clearly a situation that is not at all satisfactory from the point of view of innocent parties, particularly the airline.

Part III
Liability

The third part of this book is dedicated to the vital issue of liability. It is potentially a complex and vast subject of boundless proportions. Out of necessity, therefore, specific areas of liability are selected for review. Prominence is given to a particular relationship, that between the airline operator and its passengers and consignors of cargo during the carriage by air. It is clearly from this situation that problems of liability are most likely to arise and the relationship between the airline and its customers will be considered in some detail. This relationship has been the subject of considerable debate and regulation on an international and national level over many years.

Before we look at any specific laws in detail, the concept of liability should be addressed and set within its context in a legal framework. Many different aspects of the law can affect airline operators and other participants in the aviation industry. There are different obligations and duties owed arising out of various legal relationships that require some definition if the reader is adequately to understand how and when a liability is owed by one party to another.

24 The Concept of Legal Liability

We must first consider what exactly is meant by liability in law. There are two basic constituents of the concept of legal liability, namely, law and obligation. Legal liability can be defined, therefore, as an obligation based on and enforceable by laws applicable to a particular jurisdiction and situation. Essentially it is a legally binding obligation.

We should next consider what we mean by such an obligation being legally binding. In other words what do we mean by law. It should clearly be distinguished from general usage or figures of speech such as laws of science or reference to general principles or established standards. We are not here concerned with morality, ethics or even justice but that which is based on the law of a given state. Law is essentially, therefore, a rule of human conduct imposed and enforced upon the members of a particular state. Essential to its existence and purpose is enforceability which is, of course, carried out through the courts of a particular judicial system. In practice there must be the two elements of order and compulsion for a legal system to work and within that system one can establish what liabilities exist and how they can be protected through the judicial system.

Sources of law

What is the body of rules that effectively becomes the law of the land? The sources of UK laws are various and in England part of a fascinating historical development. In broad terms the laws of most nation states are created by a governing body and expressed by the precise written word. Mostly this is incorporated in legislation produced by government or in the context of the UK passed by Acts of Parliament. In some jurisdictions the written law is embodied in a code such as that which originates from the Roman Empire and is reflected in more recent history in the Napoleonic Code on which French law is based.

The other common, but less obvious, source of law is the body of previously decided cases broadly known in the UK as the common law. This is an essential part of the law of the UK and certain other notable jurisdictions such as North America which likewise has a system based on precedent or case law. In effect this law is continually being made and developed by the courts. In common law jurisdictions previous decisions of the courts become enforceable by reason of the doctrine of precedent which requires that in future cases on similar facts, previous decisions are followed. It is, therefore, both a self-perpetuating and constantly evolving system.

The common law is a vital part of the English legal system. Such common law systems are fundamentally different from systems where the law is based purely on written codes or statute law as in many European countries. Whilst not a topic for debate here it is, of course, an interesting point to consider in relation to future developments within the European Community. At present the legal systems within various states are fundamentally different. Any attempt to create one multi-state nation would have to address this issue which would clearly impact on aviation as an international activity.

To summarise we can see that the sources of law are various. In the UK we have a system based on a well established historical source, the common law, and the increasingly important modern source, statute law. Together these constitute the basis of English law but they are not definitive. For example, more recently, since the UK's entry into the European Economic Community (EEC), by virtue of the European Communities Act of 1972, EEC law has played an increasingly important part in its legal system. From this rich and continually evolving pattern we will see in the ensuing chapters how certain aspects of the law affect aviation.

Classification

Having established the nature of law and its various sources the next step is to categorise it by type. There are many different ways of categorising and sub-dividing such a vast volume of information. For these purposes we need bear in mind only three basic divisions:

- criminal and civil law;
- public and private law; and
- international and domestic law.

A primary distinction should be made between criminal and civil law. The former involves wrong doings, termed offences, against the state. They do not necessarily violate any private right but constitute an act of disobedience of the public law which is punishable by the state, normally by fine or imprisonment. It is the police, as public servants, whose duty it is to prevent and detect crime and ensure that offenders are prosecuted before a court of law. Theoretically, therefore, a crime is a wrong done against, and punishable by, the state.

In contrast civil law embodies the rights and duties of persons (which can include both individuals and legal entities such as companies) towards each other. Such rights and duties are enforceable between those persons through the judicial system. This body of civil law covers many different areas, for example, property, succession, family and also includes the important areas of contract and tort which will be considered in more detail in Chapters 25 and 26 as being of significance to aviation.

A further distinction should be drawn between private and public law. Private law is that which relates to the rights and obligations of persons and essentially covers the area of civil law. Public law is concerned with matters that affect the public at large and covers such areas as constitutional and administrative law as well as criminal law.

Finally, a distinction should be made between international and domestic law. International law relates to legal issues between sovereign states as opposed to domestic law within a jurisdiction of a particular nation. Rules of international law manifest themselves in international treaties and conventions and govern, for example, the bilateral agreements relating to take-off, landing and overflying rights negotiated between the governments of sovereign states. International law is often ineffective as a body of law. This results from the obvious difficulties of creating and maintaining an overriding authority capable of enforcing rights and obligations between sovereign states.

In the context of aviation we are most concerned with civil law as it manifests itself in domestic and international legislation.

25 Contract Law

Contract law affects the aviation industry insofar as most commercial operations enter into agreements or legally binding contracts in the course of their normal commercial activities.

A contract in law is no more nor less than a legally binding agreement between two or more parties. Under English law a contract can be written, verbal, or implied by conduct. There are a few exceptions to this general rule, the most notable of which is a contract for the sale or other disposition of land which must be evidenced in writing as provided by Section 40 of the Law Property Act 1925.

Contract law has evolved as part of the common law, developing to meet the needs of a particular age. In more recent times, however, it has also been created by legislation. It is important to remember that legislation, in other words laws made by Act of Parliament, will always override and take precedence over the common law. Thus, for instance, the sale of goods, employment and landlord and tenant legislation intervenes in the normal relationship between the supplier and customer, an employer and employee and a landlord and tenant. The state has intervened in these instances, and there are many more examples, to protect the basic rights of one party which may otherwise be prejudiced. This goes to the essence of contract law which presupposes the ideal of free bargaining between two or more parties able to negotiate fairly amongst themselves. Where there is felt to be an imbalance such as the bargaining position of one party overpowering that of the other, statute may intervene to redress this in an attempt to protect the rights of the weaker party. Apart from specific exceptions, however, the basic concept continues, i.e. that a contract reflects the intention of the parties as agreed between them.

Lengthy volumes have been written on the subject of contract law and all that can be attempted here is to touch on the very basic components of a legally-binding contract. In summary there are certain key elements. First, there are normally (with a few exceptions) no specific requirements as to *form*. It is clearly advisable though that parties to a contract put the terms in writing so that in the event of a dispute there is evidence as to what was agreed.

Second, the *intention* must be present to create a legally binding contract between the parties. The essential terms must be agreed and a contract in law must be distinguished from, for example, mere negtiation or a domestic arrangement, which was never intended to create a legally binding contract enforceable by the courts.

Third, in order to reach the state of contractual agreement the parties are assumed to have gone through a stage of bargaining from which their contract results. Conceptually the process proceeds through basic stages referred to as *offer and acceptance*. An offer is proposed by one party and rejected or accepted by the other. Offers can be withdrawn at any time up until acceptance, or counter-offers can be made by the other party. Once acceptance of the terms offered has been made the contract will normally become legally binding. There must be moreover a consensus *ad idem* or meeting of minds in order that agreement can be reached.

Fourth, it is essential for a simple contract (essentially one not under seal) to include *consideration*. This is normally the price that is given by one party to the other in consideration for services or goods provided. Consideration is normally money but may not always be so, nor does it have to be adequate or appropriate. For example, an oral statement by A to B that A would sell his car to B would be a mere promise or statement of intention. There is no consideration from B and, therefore, no binding contract. If, however, A and B have negotiated and A has agreed to sell his car to B for £5, that would create an enforceable contract in law (assuming the other requirements are satisfied). The consideration in this case would be B's agreement to pay the £5. Whether the agreed consideration was a peppercorn, an act or an amount of money reflecting the true value or not, is irrelevant.

Fifth, there must be *certainty* as to the main terms of the contract. Even when a number of terms are already agreed they would probably not be enforced by a court of law at that stage if any essential terms remain unagreed.

Finally, there are a number of other conditions such as genuine *consent* given without duress by each party who must have the legal *capacity* to make a contract, for example requirements as to age and mental fitness. *Performance* must be possible and legal. Certain contracts, for example, are illegal or unenforceable if they fall within established categories such as those in restraint of trade or for an illegal act.

Privity of contract

For the reader without any legal training or contractual knowledge the above is merely a starting point in understanding the bare essentials of a legal contract. Before concluding this summary the point should also be made that all the conditions mentioned above are, of course, between the parties to the contract. This may be an obvious point but one that in practice is often missed. In general terms, a contract only creates rights and duties between the parties who have entered into and agreed its terms. Conceptually this is known as the doctrine of *privity of contract*. In practice a person may be harmed or otherwise affected by a contract to which he is not a party. This third party, however, is generally unable to enforce the contract or sue by virtue of it any party to the contract, and vice versa. This, of course, goes to the root of a legal contract based on the concept that it is the result of an agreement reached by all the parties intending to be legally bound.

In complex situations of liability this can be significant. Take, for instance, a situation where an aircraft crashes killing and harming passengers. It subsequently turns out that the aircraft, which is new, crashed because of a structural defect caused by the manufacturers. A victim (or the victim's estate) would have a contractual relationship with the airline from whom he had bought a ticket but not with the manufacturer. A passenger, therefore, would normally have no right of action in contract against the manufacturer. By contrast the airline, if it acquired the aircraft direct from the manufacturer, may have (depending on the terms of a contract between them).

Incorporating terms

The contractual relationship between an airline and its customers (normally passengers or consignors of cargo) is important and should be carefully considered. In theory, the ideal is that both parties should be in a free-bargaining position in which to negotiate the terms which they will both accept. However, the airline has the opportunity to prescribe its own terms and conditions on which it will carry passengers or goods and, in reality, customers are rarely in a position to negotiate and readily enter into the contract often without full consideration of the terms they are deemed to have accepted. How often does an intending passenger go to the airline booking office to negotiate the terms of passage? One may check a few vital details such as points of departure and destination, dates and price but probably no more. In fact, one may be deemed to have accepted a whole volume of written terms and conditions of carriage by accepting the ticket.

Standard terms and conditions of contract

This is where the airline has an advantage as the supplier of the services: it can prepare standardised conditions subject to which it will contract to carry passengers who as individuals purchase tickets for a required journey normally without any thought that they are in effect negotiating and then agreeing a contract with the airline. This gives the supplier of the services an opportunity to protect itself and introduce contractual terms to its particular advantage.

How is this done? Clearly the airline cannot conduct negotiations with every customer and then draw up with each an individual contract. In practice it produces written conditions of carriage which are normally published in a small booklet and generally applicable to its services. This should be available for every potential customer to see and accept before entering into the contract for carriage, i.e. purchase of a ticket. Purchasers have, of course, no real chance of changing any of those contractual terms. In reality they either accept them with the ticket or decline the purchase. The airline in that situation, as with many large corporate suppliers of services to numerous individuals, can, so long as its market position is secure, operate on a take-it-or-leave-it basis. This is, however, subject to two important factors: first, the legal effect of the inclusion of such terms; and, second, terms and conditions implied by statute or other authorised bodies.

Before leaving the law of contract we must return to the original basic principles referred to above. A contract is in essence a collection of terms agreed between the parties. This means neither party can unilaterally introduce terms into the contract without them being accepted by the other; nor of course, can terms be introduced after the contract has been entered into without the alterations being mutually agreed.

This fundamental principle has been tested and reaffirmed by the courts. A good example is that of a public car park: a person having decided to use the car park will normally enter via an electric barrier which opens once they have acquired a ticket and after which the contract is being performed and will not terminate until the driver leaves the car park by the appropriate exit, normally by the lifting of another barrier on payment of the fee. In a particular case it was shown that the terms and conditions upon which the car park operator relied (in this instance it was trying to rely on a term excluding it from certain liabilities) were displayed in such a way that they were only visible once the customer had taken his ticket and entered the car park. The court held it could not in those circumstances enforce any of those conditions. The point is that the car park operator was notifying the customer of its terms *after* the contract had been made such that the customer had no chance to decide whether or not to accept them and enter into the contract.

The general principle is then, that all such standard terms and conditions must be fully disclosed in advance so that the prospective customer can decide whether or not to accept them. If they do, and enter into the contract, the courts will normally enforce the terms as agreed between the parties. The courts will resort to one key question: 'What was the intention of the parties?'

In an airline context, the airline, if it wishes its terms and conditions to be enforceable through the courts, must disclose them to the customer before the contract is made. This does not necessarily mean that all the terms must be produced for customers to see but they must first, be made aware that they exist and second, be given an opportunity to examine them if they wish to do so (obviously before entering the contract).

This situation, correct in law, is sadly all too seldom adhered to in practice. The average customer on purchasing an airline ticket is unaware such conditions exist, is not made aware of the fact and does not make any enquiries. Random spot checks have in the past revealed that ticket-sales personnel can be unprepared for a request to see any conditions of carriage and even if they are aware of what exists may not always be able to produce a copy for the customer's use. In such a situation the enforceability of the conditions of carriage is very much put into question. This can only be to the detriment of the airline which having gone to the trouble of producing conditions then finds that when it needs to rely on them they are not enforceable in law.

Conclusion

For terms and conditions of carriage to be enforceable in law an airline should ensure that the customer is made aware of their existence before entering into the contract and that the opportunity is available for them to see them on request.

In practice the ticket will often refer the passenger to the full terms and conditions of carriage in addition to printed terms on the ticket. Whilst this tends to be normal practice it is doubtful whether this is sufficient to satisfy the law. It will, of course, depend on the circumstances but clearly when a contract has already been made, the reservation finalised, the price paid and the ticket say, sent in the post, it is too late to then draw attention to other terms on receipt of the ticket. By that time the contract has already been made. This is an area where law and practice often do not accord. At the end of the day this could be to the detriment of the airlines. Furthermore in many situations obligations imposed upon them by statute will apply rendering contractual terms irrelevant or inappropriate. The effect of these is discussed below.

The legal effect of the terms and conditions of a contract of carriage has been dealt with here but a further discussion of the form and content of documents of carriage will be found in Chapter 30.

Implied terms

It will be noted that the source of English Law is essentially twofold: common law and statute. Where statute intervenes it takes precedence over the common law. In short, whilst parties to a contract can normally agree whatever terms they wish, in certain situations legislation has been passed to imply specific statutory terms. As already mentioned legislation has made a significant impact in such areas as employment, sale of goods and landlord and tenant. In aviation this is also the case in terms of the liability of a carrier of passengers and cargo during the carriage by air. The Carriage by Air Act 1961 (incorporating the Warsaw Convention as amended) imposes many terms which lay down the whole basis of a contract in law between the passenger or consignor of goods and the carrier. This will be looked at in more detail in Chapter 30.

Another significant example affecting the liability of the carrier is the provisions imposed by the CAA on the holders of air transport licences. Normally such licences are granted subject to a number of standard conditions including at the present time that known as 'Condition H'. This states that the licence holder 'shall enter into a special contract with every passenger to be carried under this licence' to increase the limit of the carrier's liability to a minimum stated amount. The effect of this will be reviewed when examining the carrier's liability under the Warsaw Convention in Chapter 30.

Under the general law the exclusion of liability for breaches of contract and negligence is restricted by statutory provisions, in particular the Unfair Contract Terms Act 1977. Whereas such provisions affect many contractual situations their relevance to the airline operator is limited in that other legislation covers many of the situations affecting the domestic and international operator, particularly in respect of its liability to passengers and cargo during the carriage by air. This legislation will be reviewed in the following chapters.

26 Tort

What is the nature of tortious liability?

A tort is essentially a civil wrong (the Norman-French word 'tort' meant wrong) and arises where a person is in breach of a legal duty owed to another. It should be distinguished from both a breach of contract or a crime. The former arises out of rights and duties agreed between, and enforceable by, parties to the contract. In tort, however, such rights and duties exist by virtue of the law itself and are owed to persons in general. The law of tort is, therefore, potentially far wider in its application than that of contract. It is also distinguishable from a crime which is a wrong against the state resulting in prosecution and punishment, whereas a tort is a civil wrong for which the remedy is civil damages, not imprisonment or a fine.

The general principles of tortious liability

In order to recover damages for a tort, a Plaintiff (the person bringing the action) must prove the following:

- first, that there was a legally recognisable right or duty owed in the circumstances, i.e. a duty-situation existed;
- second, that this has been breached by the act or omission of the Defendant (or some person for whom he is responsible); and
- third, that damage has occurred as a result of that act or omission.

These are general principles with particular requirements applicable to different types of tort. It should be noted that whilst proof of damage is essential to most torts there are exceptions, for example, in certain cases of trespass which are actionable *per se* without proof of damage.

There are a number of possible defences available to a Defendant (depending on the tort) to such a claim, the most obvious being that of consent, often in law referred to by the Latin phrase 'volenti non fit injuria' (meaning no injury can be done to a willing person). A practical example of this concept is where someone engages in a particularly dangerous sport or other such activity knowing that the risks of personal injury are high.

The different types of tort

Tort, like contract, is a fundamental area of law which has evolved over many years and forms part of the backbone of the English common law system. It an be categorised into a number of quite distinct areas, the most common of which are the torts of trespass, nuisance, negligence and defamation. The latter has little relevance here but each of the former three impinge to some extent on the world of aviation.

Trespass

A trespass is an unlawful interference with the person, land or goods of another and is normally actionable *per se* without proof of actual damage. This unjustifiable intrusion can include entering air space belonging to another. The logic for this stems from the historical premise of land law that the ownership of land includes not only the surface of the land but down to the centre of the earth and up to the sky, or as stated in the ancient Latin maxim 'Cujus est solum cjus est usque ad coelum et ad infernos'. The advent of air flight in the twentieth century introduced an interesting new problem. The extent and enforcement of such rights in air space had never before been properly assessed and tested. As established by the common law, it seemed that intrusion into air space above land could be a trespass, though this may not automatically be so at any height. The position has to some extent been modified by case law by, for example, enquiring whether the infringed air space is necessary for the use of the land below. This opened the door for more reasonable interpretation in relation to aircraft. It was, therefore, suggested that flight over land at a reasonable and safe height without interference was not trespass.

Recognition of the growing problem with the development of air flight during the twentieth century led inevitably to statutory intervention. Legislative measures going back some 40 years have been taken to clarify the situation. The current Civil Aviation Act 1982, Section 76 severely restricts the right to bring an action in respect of (most) trespass aircraft. Section 76 (1) states

> No action shall lie in respect of trespass or in respect of nuisance, by reason only of the flight of an aircraft over any property at a height above the ground which, having regard to the wind, weather and all the circumstances of the case is reasonable, or the ordinary incidents of such flight, so long as the provisions of any Air Navigation Order and of any orders under Section 62 above have been duly complied with and there has been no breach of Section 81 below.

This effectively removes the right of action in respect of trespass (and nuisance as discussed later) in normal circumstances. It produces what must be a practical solution to the problem produced by scientific developments challenging ancient law. The solution, however, is not absolute and it is clear from the Act that the rights of such aircraft are limited by certain qualifications. The existence of the test of reasonableness is in itself an indecisive limitation, but the Act tries to reach a practical and fair solution. It does not totally abandon the rights of persons on the ground. Section 76 (2) deals, in particular, with surface damage.

> ... where material loss or damage is caused to any person or property on land or water by, or by a person in, or an article, animal or person falling from, an aircraft while in flight, taking off or landing, then unless the loss or damage was caused or contributed to by the negligence of the person by whom it was suffered, damages in respect of the loss or damage shall be recoverable without proof of negligence or intention or other cause of action, as if the loss or damage had been caused by the wilful act, neglect, or default of the owner of the aircraft.

Section 76 (3) provides an indemnity to the owner of the aircraft in certain circumstances where another person is legally liable to pay damages in respect of material loss or damage as aforesaid. Further protection is given to an owner in Section 76 (4) in certain circumstances where an aircraft has been demised, let or hired for more than a minimum period.

The liability for surface damage is absolute in the circumstances stated in Section 76 (2) except where the plaintiff is contributorily negligent. This concept of absolute liability is known in law as *strict liability* and is often introduced by statute. There will be further examples of this when reviewing the liability of the carrier during the carriage by air. Strict liability confers an automatic liability, without requiring proof of fault. For example here, an aircraft may suffer an unfortunate technical problem such as a piece of the wing falling off onto the ground below, seriously damaging a house and injuring an inhabitant. Even though it may be a genuine accident with no fault on the part of the owner of the aircraft, the owner is still prima facie liable under Section 76 (2) for loss and damage caused.

Section 81 deals with dangerous flying and is a further qualification to Section 76 (1). It provides for the pilot or person in charge of an aircraft, and potentially the owner, to be punished by fine or imprisonment where an aircraft 'is flown in such a manner as to be the cause of unnecessary danger to any person or property on land or water'. The owner (which may include a hirer) has a defence if it can be proven to the court that the aircraft was flown 'without his actual fault or privity'. Restrictions on aerial advertising are also prescribed by this part of the Act.

These provisions relating to harm caused to persons and property on the ground are by no means comprehensive and may involve a variety of other legal considerations.

Nuisance

In general terms a nuisance is an act or omission which interferes with the use or enjoyment of land of another. It can be distinguished from a trespass to land in that the interference need not be direct or involve physical entry on the land: for nuisance there must be proof of damage, while trespass is actionable *per se*.

There are two basic types of nuisance: private nuisance and public nuisance. Private nuisance is essentially an unlawful interference with an individual's use, or enjoyment of their land or property. It is a wrongful act or omission which must cause harm. This may affect their health, comfort, or convenience. The most common examples are neighbour disputes involving noise, smoke or damp affecting a next-door property. A public nuisance is also an unlawful interference, but with a public right which affects the public generally or some section of it. Aircraft noise or interference is a particular example of a public nuisance.

As we have seen from looking at the tort of trespass, a legislative solution has been provided to try and resolve the problem of aircraft interference with the use and enjoyment of land below. The reader is referred to Section 76 (1) of the Civil Aviation Act 1982 quoted on p. 149. This makes it clear that Section 76 is applicable to nuisance as well as trespass.

The issue of aircraft noise is dealt with in some detail in part IV of this book on environmental matters, in particular, statutory provisions referred to in the Civil Aviation Act 1982 are discussed on pp. 236–8. Suffice it to say here that the statutory removal of the right, in normal circumstances, to sue for noise nuisance from aircraft has been the subject of increasing criticism. In effect, it denies the public a vital legal weapon in opposing aircraft noise. Vocal protest and political lobbying are now the only practical avenues open to the growing band of noise opponents. It should be added, however, that as will be discussed in Part IV, the British government has over recent years introduced increasingly strict controls on aircraft noise.

Negligence

The tort of negligence has steadily developed in significance, application and extent. Negligence is the breach of a duty to take care which results in damage to the plaintiff by the defendant. There are three basic components which must be present in order to establish negligence:

- the defendant must owe the plaintiff a *duty of care*;
- there must have have been a *breach* of that duty;
- that breach must have resulted in *damage*.

If there is only one case that the non-lawyer reader should be referred to it is that of *Donoghue* v. *Stevenson* [1932] AC. 562. It is a leading case in the development of negligence and must be one of the best known of all cases. In summary, it was alleged that Mrs Donoghue drank some ginger beer out of a bottle given to her by a friend who had purchased the drink. After Mrs Donoghue had consumed part of the bottle she found that there was a decomposed snail inside it. It was alleged Mrs Donoghue later became ill, due to drinking the ginger beer tainted with the dead snail. The significant legal issue was the fact that the Plaintiff who suffered damage, namely Mrs Donoghue, was suing the manufacturer with whom she had no contractual relationship whatsoever. Her only claim was thus in tort, on the basis that the manufacturer of the ginger beer owed her, as an ultimate consumer of his product, a duty to take care in the production of that product. In an all-important speech by Lord Atkin in the House of Lords it was established that a manufacturer did in such a situation owe a duty of care to an eventual consumer. In describing the circumstances in which a duty of care arose, Lord Atkin introduced the neighbour principle which has since assumed the role of a fundamental legal principle in assessing where a duty of care is owed. One cannot do better than quote the actual words of Lord Atkin in his memorable speech:

> You must take reasonable care to avoid acts or omissions which you can reasonably foresee would be likely to injure your neighbour. Who, then, in law is my neighbour? The answer seems to be persons who are so closely and directly affected by my act that I ought reasonably to have them in contemplation as being so affected when I am directing my mind to the acts or omissions which are called in question.

The relevance of negligence to aviation is clearly limitless. With many thousands of people giving and receiving services which involve the use of complex and highly technical machinery there are potentially manifold situations in which a duty of care is owed. Obvious examples involve the manufacturers of aircraft and equipment and airlines who fly and use them. An aircraft manufacturer would normally owe a duty of care to the people who fly in and use that aircraft. If a construction defect causes a new aircraft to crash resulting in death and injury, those harmed may well have a valid claim in negligence against the manufacturer. Of course, this would depend on the plaintiffs proving the basic principles of negligence: that the defendant owed them a duty of care which had been breached and directly caused the damage alleged.

One important consideration is that of remoteness of damage. The damage caused must be a reasonably direct result of the defendant's act or omission. A defendant is not necessarily liable for all damage flowing from his negligent act or omission, however unlikely, but only for such damage as was reasonably forseeable.

Airlines owe a duty of care to their passengers and crew in a number of ways. An air disaster is the obvious situation in which negligence would be considered if there appears to be any fault on the part of the operator. A major air disaster could become a legal quagmire of claims and international conflicts of jurisdiction, such that from the early days of flight nations have co-operated to produce and effect an international legal code adopted by participating nations. The various conferences held, originating with the famous Warsaw Convention, have resulted in many nations adopting standardised laws in relation to a carrier's liability to its passengers and consignors of cargo during international carriage. This will be examined in some detail in forthcoming chapters.

So, once again we see an example of the common law position being replaced by statutory intervention, namely, in certain vital areas of potential negligence claims arising out of carriage by air, legislation has laid down the rights, liabilities, and remedies of the parties. The legislation is applicable, however, only to specified situations and any potential claims not covered would rely on the ordinary law of negligence.

One such area is that of people affected by an accident, though not actually involved. For example, relatives who have suffered severe nervous shock on the news of the death or injury to a loved one in an accident. This may be exacerbated by various levels of proximity from actually watching the accident happen, to hearing it some distance away, seeing it on television or just the delivery of the news itself. Many jurisdictions would not entertain such a claim, and whilst a relatively new area of law in Britain, it is continually developing to widen the scope of possible plaintiffs. For instance, in some accident cases it has now been held that not only could families of victims several miles away claim damages, but also those rescuing victims. This is, of course, always subject to their proving the essentials of negligence in that the duty of care had been breached thereby causing them harm. A major determinant of whether or not a duty of care is owed is the extent to which the defendant ought to have foreseen that a negligent act or omission on his part might cause damage to the plaintiff. The test of foreseeability is really one of extent, and it is easy to see how the goal posts can be extended as the players increase. When a major disaster occurs, such as an aircraft crash, many people could be affected. From those in the aircraft who suffered death or injury the net could widen to their relatives hearing the news, to independent persons involved in the rescue operations, and to onlookers shocked by the sight of the accident and its victims.

Conceptually this has been referred to as the 'zone of danger' test. The zone has probably been stretched farthest in North America which tends to lead the way in widening the interpretation of such common law claims.

27 Product Liability

Before leaving this general review of legal liability some attention should be given to claims arising out of product liability. This is another developing area of the law and is concerned with claims arising out of a faulty product.

Looking at a particular type of problem one can see how various different legal remedies could be applied. If the product leaves the factory with a defect that then causes damage, a potential litigator would need to consider on what basis it could make a claim in law. So far, we have looked at different areas of the law to establish some of the basic legal rights and remedies that exist. In now looking at a type of situation we can see how different laws could be applicable to the facts of a particular case.

We may consider, for example, an aircraft accident, similar to that of the British Midland Boeing-737 aircraft that crashed onto the M1 motorway near East Midlands Airport at Castle Donington, Leicestershire, England, in 1989. Significantly, the aircraft was relatively new, encouraging speculation about a possible manufacturing defect. The aircraft appeared to have crashed due to engine failure, the reasons for which were not at all clear. It was reported that there had been a fire in one engine as a result of which an engine had been shut down. There was then speculation as to whether this was the correct engine, as the surviving engine had then failed shortly before the aircraft crashed on its final approach to the airport for an emergency landing.

There followed months of investigation and speculation about the originating cause of the crash and where liability lay. Subsequently, reports have been made and certain conclusions drawn which are not a matter for consideration here. Leaving aside the actual case and using the crash merely as a hypothetical example, we can consider various questions of liability that could arise.

In considering the position of potential claimants, the law will only provide a possible remedy in terms of compensation if there exists a right of action: in other words, has some law been infringed or breached?

We have seen that in general terms we have two main sources of law: the common law and statute law. Leaving the second aside for the moment let us look at the first. We have seen that there are two particularly important areas of the common law, contract and tort. Breach of contract provides the harmed party with a potential cause of action. There must exist a legally binding, and therefore enforceable, contract. It will, of course, only cover the terms actually agreed between the parties, and more significantly, normally only those persons who are party to the contract are legally bound and in a position to enforce its terms. This, as we discussed before, is known as the doctrine of privity of contract which is, in practice, a limiting factor as to who may sue. For instance, where an aircraft crashes causing death and injury which is subsequently proved to have been caused by a defect in the manufacture of the aircraft for which a particular manufacturer is responsible, there may be a variety of possible claims depending on the claimant's legal relationship with the manufacturer. From a contractual point of view, any prospective plaintiff would have to have entered into a contract with the manufacturer, the terms of which had been breached. Thus in the case of passengers and other third parties, such as owners and occupiers of property on the ground, it would be highly unlikely that they would have a right of action in contract.

The case of the airline operator is, however, more hopeful. It would presumably have entered into a contract to buy the aircraft in the first place. This would probably contain warranties about the condition of the aircraft that, depending on the terminology of the contract and the facts of the case, might cover the defects. It should also be remembered that, as already noted, there has been considerable statutory intervention into the common law which may mean extra terms are implied by statute into the contract.

In conclusion, the doctrine of privity greatly limits the scope of claims for breach of contract. Most injured parties would be looking at the tort of negligence for their legal remedy. In tort, as we have seen, the rights and duties already exist in law and apply to the public in general, whereas in contract rights and duties are created voluntarily by certain persons who alone can enforce them. Tortious liability protects third parties in contrast to contract law.

In our example, therefore, injured parties may have a right of action in law against the aircraft manufacturer for negligence. It may also be that the operator of the aircraft would have a right of action against the manufacturer for both breach of contract and negligence. All remedies would, of course, depend on the legal requirements having been met.

The common law duty of a manufacturer to a consumer of its product was established, as we have seen, by the case of *Donoghue* v. *Stevenson*. This area of law has since expanded with the consumer boom of the twentieth century. The most dramatic changes in Britain have been made by recent statutory intervention. These moves have been provoked by an EEC Directive of July 1985. The debate in the United States and Britain had in recent years moved towards imposing strict liability in the area of product defects. In general terms, the EEC decided that this was the best solution, as stated in the Council Directive of 25 July 1985 (No. 85/374/EEC):

Liability without fault on the part of the producer is the sole means of adequately solving the problem, peculiar to our age of increasing technicality, of a fair apportionment of the risks inherent in modern technological production.

The Directive essentially provides that a *producer* be liable for *damage* caused by a defect in its *product*. Those three key terms are defined by the Directive in order to clarify the implementation of the principle. Damage, not surprisingly, includes death, personal injury, and certain cases of damage to property. Product significantly includes most moveables and, for example, electricity. More surprising is the definition of producer which widens the interpretation that might ordinarily be used. It includes not only the manufacturer of a finished product but also, for example, certain others involved in the creation of the product such as producers of raw materials or component parts, and even importers of the product into the EEC for sale, hire, lease or other distribution in the course of their business.

The prospective plaintiff does not, therefore, have to prove that a duty of care exists between himself and the producer, but merely that he has suffered damage caused by a defect in the product. Few defences are afforded the producer, though it is given some necessary protection, for example, where the product was not put into circulation (e.g. a test flight) or where the defect which caused the damage did not exist at the time the product was put into circulation.

Significantly no maximum limit of liability was prescribed (unlike the Warsaw Convention, see Chapter 30), but there were provisions enabling individual member states to limit certain death or injury cases to an amount expressed as not less than 70 million Ecus. This gives an indication of the general principles laid down by the Directive, the text of which should be consulted for a full and precise understanding of its terms. The next step was for the British and other EEC governments to put into effect the EEC Directive by national legislation.

The result in Britain is the Consumer Protection Act 1987, Part I of which essentially gives effect to the EEC Directive and came into force on 1 March 1988. The Act follows the principles of the Directive and imposes strict liability in respect of certain types of damage caused by a product defect. Damage includes death, injury, or damage or loss to property (with certain qualifications). The product is widely defined to include any goods or electricity and a component part or raw materials. The producer is equally widely defined being not only the person who manufactures the product but also other defined categories of person. The defences are limited and the Act, which is something of a legal watershed, should be reviewed with care by all those who may fall within its net of liability.

28 Carriage by Air

We have considered some of the recent developments in product liability and how these could affect the manufacturer or producer in the aviation industry. We turn now to the area of most concern to the airline operator. Most accidents causing injury and damage occur whilst passengers or goods are being transported in an aircraft. We have seen that both the law of contract and tort (particularly the tort of negligence) could be relevant, as the relationship between an airline operator and its passengers and consignors of cargo would normally be both contractual and one in which a duty of care was owed. However, legislation has intervened in a number of areas, and in particular, in relation to the rights and liabilities arising out of the carriage by air as between the airline (or 'carrier' as defined by the legislation) and its passengers and consignors of cargo.

These rights and liabilities are in force by virtue of the Carriage by Air Act, 1961 and are limited to a period of time during which the carriage by air takes place. We will see though that in practice it has not always been easy or possible to apply the theoretical limits at a precise moment in time and this can be a matter of great importance to the parties involved. Depending upon whether or not a claim falls within the provisions of the Act could substantially affect the issue of liability and any compensation or damages received. We will examine the statutory provisions in some detail in the following chapters, but two essential differences are worth noting now.

First, the legislation (with certain qualifications) imposes strict liability on an airline for death, injury and damage caused by an accident during the carriage by air to persons, baggage or cargo. As in the product liability legislation reviewed in the previous chapter, the plaintiff would not, therefore, have to prove negligence or, therefore, that there was a duty of care that had been breached.

Second, there are fixed monetary limits on the liability of the carrier the value of which has in relative terms become increasingly low over the years.

In conclusion, therefore, a plaintiff would probably find it far easier to prove a claim under the legislation. Indeed, it would be far more likely that the plaintiff would be saved the necessity of a court action as the claim and its remedy would be provided for by statute with less likelihood of argument. However, the monetary amount of damages to which the plaintiff could be entitled may well, particularly as against a substantial claim in negligence, be much lower when subject to the statutory limitations on the carrier's liability. The carrier in this situation may prefer to be subject to the legislation rather than chance a negligence action. This will, however, depend on the facts of the case and, in particular, the size and merits of the claim.

Before we review that legislation we will need to see how it arose and the problems it tried to solve. The following chapters will deal with this vital area of potential liability for the airline operator, dealing first with international carriage, dominated by the Warsaw System, and then domestic carriage, with its separate but not dissimilar provisions.

29 International Carriage: the Warsaw Convention

Predicament

It was not long after international flights were first undertaken in the early decades of this century that it was recognised that the legal problems arising from any such 'international' accidents would be complex and even irresolvable. Technology had advanced into a new era which crossed the boundaries not only of countries and air space but also national jurisdictions and legal systems.

The problem was essentially that of *jurisdiction*. Exactly what law was to apply to whom and where? How could you reconcile the different international rights and laws imposed and enforced by quite different and autonomous legal systems? Any investigation produced the ultimate legal dilemma – *a conflict of laws* situation.

In practice what does all this mean? Let us consider in practical terms the following simple example:

A passenger called 'P'
Is a national of country 'C' = Passenger nationality
Buys an airline ticket in country 'T' = Place of contract
To his destination country 'D' = Place of destination
Aboard an airline from country 'A' = Airline nationality
'P's luggage is lost in country 'L' = Place of breach

What is 'P' to do? Does he bring an action, where he lives or where he bought the ticket, where the airline is based, or where it breached the contract, or where he was flying to and made the complaint, or where the luggage was lost?

Add to this the different situations of every passenger on board an international flight. There could be many different nationalities all having bought tickets from different places and even airlines perhaps flying different sectors. If the aircraft crashes what claims could be brought and where? Even if a court decides it has jurisdiction to hear a particular claim, how will it be able to enforce a judgment it may make against a foreign defendant airline? Furthermore, the plethora of different claims in different jurisdictions will mean a complete lack of uniformity or predictability for airlines, passengers, or insurers. It will fail to provide a system of justice to the people in general on that aircraft. In conclusion, the result would be *legal chaos*: a complex situation of conflict and confusion.

Solution

The ideal solution has to be a uniform system of laws governing international carriage that is imposed upon and enforced by all nations involved in international air travel. There is also the fundamental problem of enforcement in international law. Given that most nations are independent, autonomous states, what overall authority is there to enforce such laws? As we saw in Chapter 24, when discussing the essence of legal liability, without enforcement procedures a system of law becomes meaningless. Without an overriding authority one can only rely on voluntary co-operation. To an extent this is what has happened in practice.

It was indicated in commencing this discussion that awareness of the problem has existed almost since its inception. This is indeed so and the consequence must be one of the first and most surprising international achievements this century. The element of surprise is really that of time. It was 1929 when representatives of many nations met at an international conference in Warsaw, Poland, to consider the legal problems of international carriage by air. Previously there had been various studies and discussions, particularly those by the Comité International Technique d'Experts Juridiques Aeriens (CITEJA). Their object had been to put an end to the conflict of law problems inherent in international carriage by air. It was under their auspices that the Warsaw Conference or Conference Internationale de Drôit Privé Aeriens was held. This resulted in a Treaty for the Unification of Certain Rules Relating to the International Carriage by Air, otherwise known as the *Warsaw Convention*. The Warsaw System, as it has since developed, has dominated the legal arena of air law ever since.

The Warsaw System

The Warsaw Convention took place in October 1929. The original text is in French and it was up to participating nations to incorporate the Convention in their respective statute books. The procedure for participating nations was threefold. The first stage was for those attending the Conference to sign the Convention. The second stage, known as ratification, is the adoption of the Convention by individual nations. Finally, there is a stage when the Convention is actually put into force in the respective jurisdictions.

The United Kingdom was one of the original signatories in October 1929. The Convention was then ratified in the UK in February 1933 and came into force in May of that year. Many other European countries were also original signatories as were many nations throughout the world. In subsequent years additional nations have ratified and brought into force the Convention; thus joining the Warsaw System.

Not only have the participators increased in numbers over the years, but the system itself has also developed to incorporate new changes and amendments. Further conventions were held, most notably at The Hague, resulting in the Hague Protocol of 1955, also the Guadalajara Convention of 1961, the Guatemala Protocol of 1971, and the Montreal Additional Protocols of 1975. Each of these will be considered in due course but for our present purposes the Warsaw Convention of 1929, as amended by the Hague Protocol of 1955, will be reviewed. The reason for this is that most of the participating nations have adopted both, with the most prominent exception being that of the USA which signed the Hague Protocol (but did not adopt it for reasons that will be reviewed in Chapter 31). The participation and adoption of the additional Conventions has proved to be full of problems and few are in force.

The UK signed the Hague Protocol in March 1956 and thereafter adopted and enforced it in 1967. The current legislation incorporating the Warsaw Convention, amended by the Hague Protocol, is the Carriage by Air Act 1961. The text will be referred to as the 'Convention' and will be examined in some detail in the following chapter. Further changes to the Convention and other legislative changes in the United Kingdom will be considered after a detailed examination of the Convention.

After the Hague Protocol, the Warsaw System became dogged with problems and there is little uniformity in adoption of further amendments. The years after 1955 are a subject in themselves. Suffice it to say for now, that the Warsaw Convention has been adopted by most developed nations, though not all have adopted the Hague Protocol. The relevance of this and the changes made will be drawn to the reader's attention in Chapters 32–5.

The Carriage by Air Act 1961

The purpose of this Act, as stated in the preamble, is to 'give effect to the Convention concerning international carriage by air known as "The Warsaw Convention as amended at The Hague, 1955" to enable the rules contained in that Convention to be applied'. The Act itself contains first a number of sections dealing with general preliminary matters after which follows a Schedule consisting of two parts; being the Convention in the English and French texts. The body of the Act together with the English text of the Convention is reproduced in Appendix I of this book. For proper compliance reference should always be made to the actual text of the Act.

Preliminary sections

The Act begins with a number of general points which may be summarised as follows:

- In the event of any inconsistency between the English and French texts, the French shall prevail (see Section 1 (2)).
- The Convention is effective in the United Kingdom irrespective of the nationality of the aircraft (see Section 1 (1)).
- Orders in council will from time to time certify the parties to the Convention referred to as High Contracting Parties (see Section 2).
- The limitation of liability expressed in the Convention (under Article 22) shall apply to all proceedings brought in the UK (and may even take account of proceedings outside the UK). (The effect of this is important. If it were possible for a passenger to be awarded additional damages in excess of those under the Convention, the statutory limit of liability so given would become a nonsense in practice) (see Section 4).
- References to 'damage' in Article 26(2) of the Convention shall include loss of part of the baggage or cargo in question (see Section 4A, added by Carriage by Air and Road Act 1979).
- Military aircraft and purposes are normally excluded (see Section 7).

Organisation of the Convention

The Convention is subdivided into five chapters dealing with the following:

- Chapter 1 – Applicability and Definitions
- Chapter 2 – Documents of Carriage
- Chapter 3 – Liability of the Carrier
- Chapter 4 – Combined Carriage
- Chapter 5 – General and Final Provisions

A detailed review of each chapter will follow and should be read with reference to the actual text (see Appendix I, p. 273).

Applicability and definitions

Crucial to the application of the specific rules of the Convention are the definitions of applicability. In the facts of a particular case does the Convention apply? Article 1 commences with what appears to be a relatively simple definition: 'The Convention applies to all *international carriage* of *persons, baggage* or *cargo* performed by *aircraft* for *reward*' (emphasis added). There are a number of key words within that definition. The subject matter of persons, baggage, or cargo is relatively obvious. 'Aircraft for reward' should be considered further.

There is no definition of 'aircraft' within the Act. In case of doubt it may be appropriate to consider the classification of 'aircraft' in the Air Navigation Order 1989 (see in particular Section 109 and Schedule 1 of the Air Navigation Order 1989, SI 1989 No. 2004). This, with various qualifications, makes provisions for balloons, airships, gliders, gyroplanes, and helicopters.

'Reward' also remains undefined in the Act. There is, however, a useful definition in the Civil Aviation Act 1982 (Section 105 (1)): '"Reward" in relation to a flight, includes any form of consideration received or to be received wholly or partly in connection with the flight irrespective of the person by whom or to whom the consideration has been or is to be given.' By the words 'any form of consideration' one can assume that non-monetary consideration is included.

There is a further important qualification following the initial definition of applicability. The Convention goes on to state that it shall also apply to 'gratuitous carriage by aircraft performed by an air transport undertaking'. This in practice means that the Convention applies to an aircraft undertaking irrespective of whether or not the carriage is performed for reward. An air transport undertaking is not defined in the Act but reference again to the Air Navigation Order 1989 (see Section 106) supplies a useful interpretation as 'an undertaking whose business includes the carriage by air of passengers or cargo for valuable consideration'. Again where there is no precise definition in the Act reference to similar authorities, whilst not necessarily binding, may be useful and indeed persuasive.

As we have already seen, whether a plaintiff's claim falls under the Convention or not could make a considerable difference to the success of its case, and the amount of damages recoverable. If the claim is outside the Convention, the plaintiff would have a greater burden of proof in establishing negligence, but potential damages are limitless. Under the Convention, there is no need to prove negligence but relatively low limits are set on the damages that can be gained. Thus, for carriers it is normally far preferable to be covered by the Convention and, therefore, to be protected by the limitation on liability. When considering the position of the carrier we must not purely have in mind a large commercial airline. There are many different types of flying operations from large monolithic flag carriers to small air-taxi operators and, of course, ordinary private individuals. Problems can arise with small commercial operators or even private non-commercial pilots who are unaware of, or not properly protected in terms of their legal liabilities.

It is clear from the first definitions of applicability of the Convention that a private pilot, say, taking friends on a day trip to France could find himself (perhaps quite unintentionally) covered by the Convention. He is clearly not an air transport undertaking, but is he performing international carriage for reward? If he charges his friends for the trip then he clearly is. Suppose he does not deliberately charge a fee for the trip but accepts a casual offer by a friend of a small sum to help with the expenses of the flight? Probably quite unwittingly, neither would realise the potential legal consequences of that innocent gesture which could fall within the definition of 'reward'. Supposing there is subsequently an accident in which the friend is seriously injured in a crash landing. As far as the 'friend' is concerned, carriage may be subject to the Convention, and though he may be seriously injured his claim is limited to the relatively low limits. Other friends on the same aircraft who gave nothing in terms of consideration or remuneration might be able to make claims in negligence which, if successful, could produce far higher damages for their injuries.

These discrepancies and the consequences of the various legal relationships are not always fully understood and appreciated by the parties involved. It should also be remembered that quite apart from these particular liability considerations the issues may involve other problems of a regulatory or insurance nature. For example, a private pilot may be breaching the terms of his insurance policy and committing regulatory breaches if he accepts consideration from passengers and in effect is operating, unofficially, a commercial operation. In short, every situation has to be examined as to its facts to see if and how the Convention would apply and what effect this will have on individual parties.

International carriage

Fundamental to the initial definition of applicability discussed above is the exact interpretation of 'international carriage' which is given a precise meaning constructed as a formula with two parts.

International carriage (see Article 1 (2)) is carriage where, according to the agreement between the parties, the places of departure and destination are either:

(1) both situated in the territories of two High Contracting Parties (whether or not there is a break in the carriage); or
(2) both situated within the territory of a single High Contracting Party if there is an agreed stopping place within another state (even if that state is not a High Contracting Party).

The first statement is perhaps obvious, in that there must be carriage between two nations to be essentially international and it is qualified by both the places of departure and destination being territories of High Contracting Parties. The second definition is more technical. There must be two different nations or states involved to be international, but here only *one* need be the territory of a High Contracting Party (if the places of departure and destination are both within that territory) subject to there being an *agreed* stopping place in another state (whether or not that be a High Contracting Party).

For further clarification the Convention states the converse of this second definition. Where carriage is between two points in the territory of a single High Contracting Party, namely, *without an agreed stopping place within the territory of another state* it is not international carriage for the purpose of the Convention.

It should be noted that the reference to a High Contracting Party, (being a party to the Convention) normally includes all the territories which are part of that sovereign state (see p. 199). For example, in the UK, this would normally include territories such as Barbados, Gibraltar and Hong Kong. Thus whilst travel between, say, London and Hong Kong is clearly 'international', in the ordinary sense of the word, it is not international carriage as defined in the Convention. For, without an agreed stopping place, the journey from London to Hong Kong has the points of departure and destination both within the territory of one contracting state. Such routes are often referred to as cabotage routes in the aviation industry.

There have been varying definitions of 'cabotage'. A useful definition is that provided by the CAA: 'carriage, other than domestic carriage, intended by the operator of the aircraft to begin and end at places each of which is in the United Kingdom, a relevant overseas territory or an associated state' (see the CAA's *Official Record*, Air Transport Licensing, Series 1, definition section). Note the distinction between cabotage and domestic carriage.

The applicability of the Convention with all its significant consequences depends on the precise interpretation of certain words within the given definition. For further clarification certain words or phrases are considered further below.

'Agreement between the Parties'
Obviously the definition of international carriage cannot be applied without

knowing exactly the places of departure and destination. These are to be those agreed in advance between the parties in the contract of carriage. As such, they would normally be stated on the passenger ticket. In the event, however, they may not be the actual places of departure or destination: bad weather conditions could cause a flight to be diverted to or from another airport, maybe in another country; an emergency landing could bring the aircraft down in a country not originally intended. The Convention, however, makes it clear that it is the places in the original contract of carriage that are relevant, not necessarily what may actually happen in practice.

To illustrate the point one could consider a flight between London and Hong Kong. As Hong Kong is a cabotage route and not an independent High Contracting Party, the carriage between the two points would not be international carriage, as defined by the Convention. If bad weather conditions force the aircraft to be diverted and land at Singapore, it would, in fact, be in the territory of another High Contracting Party. That would not, however, be relevant as the agreed place of destination was Hong Kong, irrespective of whether the carriage by air ended there or not.

Finally, it should be noted that the 'parties' to the 'agreement' for the carriage may not always be the actual passengers. Those who made the agreement with the airline could, for instance, be an employer for employees, or parents for a child.

'Break'

This could be anything from a brief stop-off for refuelling to a disembarking break of several weeks. This also has a bearing on what actually constitutes the contract of carriage where the passenger breaks the journey and then returns to an ultimate destination. This is discussed more fully below.

'Agreed stopping place'

This is similarly construed to include a variety of situations. The stopover may be brief or for longer, as in the case of *Collins* v. *British Airways Board* (discussed below) where the passenger had several weeks' stopover in a contract of carriage for a number of different sectors. The essential word, however, is 'agreed'. Thus, in interpreting the second part of the definition of international carriage, the stopping place, irrespective of its purpose, must be that which was agreed in advance between the parties and will not include last minute diversions or emergency and unscheduled stops.

At all times we are reminded that the carriage by air creates a contract the terms of which will include the places of departure and destination and any stopping places contemplated by the parties thereto at the time the contract was entered into.

Carriage by air

Finally, we cannot leave this vital section defining international carriage and the applicability of the Convention without clarifying one further point. The definition as so far discussed may seem clear in theory but it soon proves difficult to apply in practice. The main confusion lies in assessing the length of the contract for the carriage by air. In practice, of course, most passengers do not buy a ticket for one sector, but for two or more. Most traffic is after all a two-way journey: a flight out to a particular destination where the passenger embarks for the period of stay for holiday, business, or whatever purpose and then returns to the original place of departure. The question soon arose as to whether the outbound flight should be treated as a separate contract of carriage from the inbound flight. Where did one contract of carriage end and where did another begin? How should the terms 'break' and 'agreed stopping place' be construed in this context? Whether a return journey or a multi-sector routing was construed as *one* contract of carriage or several could have significant consequences as to the applicability of the Convention.

Consider the following example based on the facts of the case decided by the English Court of Appeal in 1937. The case, *Grein* v. *Imperial Airways Limited* [1937] 1 KB 50, (CA) involved a passenger who flew from London via Brussels to Antwerp and then returned via Brussels to London. It is essential to note that at that time Belgium was not a High Contracting Party (it did not ratify and put into force the Convention until 1936). If two contracts of carriage were to be construed between London (with the United Kingdom a High Contracting Party) and Brussels (with Belgium not a High Contracting Party at that time) neither the outbound nor the inbound journey would be within the Convention. If, however, the whole routing is taken as one contract of carriage, it falls within the Convention: both points of departure and destination are within the territory of a single High Contracting Party, i.e. the United Kingdom, and there is an agreed stopping place, i.e. Brussels, within the territory of another state (irrespective of whether or not that state is a High Contracting Party).

The Court of Appeal held (by a majority) that there was one contract of carriage (irrespective of the different sectors). The contract was 'a unit not a journey' and Lord Green concluded:

> The rules are rules relating not to journeys, not to flights, not to parts of journeys, but to carriage performed under one ... contract of carriage. The contract ... is, so to speak, the unit to which attention is to be paid in considering whether the carriage to be performed under it is international or not.

This case clarified a vital point of potential confusion and has since been reaffirmed as in the more recent case of *Collins* v. *British Airways Board* [1982] All ER 302. This latter case involved a number of sectors, including a domestic flight: Manchester–London–Los Angeles–New York–Manchester. The court held categorically that this represented *one* single contract of carriage despite the various different sectors, and irrespective of the fact that one, Manchester–London, was a domestic sector. Thus, an accident on that particular domestic sector would not prevent the applicability of the Convention because the overall contract was for international carriage.

It has also been established that a single contract can be evidenced by two or more tickets. Thus, a ticket is not conclusive evidence of the whole contract of carriage, and the carrier can not avoid or otherwise alter the applicability of the Convention for these purposes by issuing separate tickets. For instance, this was endorsed in an American case *Petrire* v. *Spantax S.A.* (1985) US Ct. of Apps 2nd Circ. The significant information is the intention of the parties when entering into the contract.

Successive carriage

This provides for successive carriage. If the parties regard the carriage as a single operation it is to be regarded as one undivided carriage even if performed by successive carriers, thus basically bringing it within the Convention. Significantly it expressly confirms the position that carriage does not cease to be international for the purposes of the Convention merely because one or more parts of the carriage are for flights within the same state.

In practice, of course, where there are a number of different flight sectors within a contract of carriage it is quite possible that two or more airlines will be involved. What is essential is that the parties agree at the onset that the carriage will be performed by successive carriers. If subsequently another carrier is substituted this is, therefore, not likely to be successive carriage within the Convention. The various practical circumstances do, however, render this open to interpretation according to the facts of the particular case. For instance, where the substitution was made with the agreement of the parties it may be accepted as successive carriage. There will, however, be no successive carriage where the contract specified a single carrier and without agreement of the other parties, that carrier substitutes another carrier for all or part of the carriage.

Excluded carriage

Article 2 deals with two miscellaneous matters. It basically includes carriage performed by the state or by legally constituted public bodies, but specifically excludes the application of the Convention to mail and postal packages. In summary, therefore, the following categories of carriage will normally be exempted from the Convention:

(1) non-international carriage as defined (Article 1 (1) and (2));
(2) gratuitous carriage where not performed by an air transport undertaking (Article 1 (1));
(3) carriage of mail or postal packets (Article 2 (2));
(4) carriage for military purposes (see Section 7 of the Act).

Documents of carriage

This chapter deals with the important area of documentation as it relates to the three categories of items carried, namely: persons, baggage and cargo. The chapter is, therefore, divided into three sections:

● Passenger tickets
● Baggage check
● Air waybill

Similar provisions relate to each section, by providing that documentation giving specified basic information must be produced before the carriage by air. As will be seen, the consequences of not complying strictly with these provisions can be serious, especially for a carrier who can lose the protection of the limitation of its liability.

Passenger tickets

Article 3 (1) states that a ticket must be delivered containing the following specific information:

(a) an indication of the places of departure and destination;
(b) if the places of departure and destination are within the territory of a single High Contracting Party, one or more agreed stopping places being within the territory of another State, an indication of at least one such stopping place;
(c) a notice to the effect that, if the passenger's journey involves an ultimate destination or stop in a country other than the country of departure, the Warsaw Convention may be applicable and that the Convention governs and in most cases limits the liability of carriers for death or personal injury and in respect of loss of or damage to baggage.

This vital information is really twofold: items (a) and (b) relate to the definition of international carriage, and (c) is the notification to passengers generally referred to as the *Warsaw Notice*.

Article 3 (2) goes on to state that the ticket is prima facie evidence of the conclusion and conditions of the contract of carriage. As we have already seen, however, the terms of the contract of carriage between the parties thereto are as they intend and agree, and the ticket may express any part of the contract. When we considered the law of contract in Chapter 25 it was noted that apart from the statutory terms parties could under the common law agree additional terms (so long as these were not in breach of any statutory provisions). It is, therefore, not unusual for airlines to incorporate additional, and often lengthy conditions of carriage into the contract of carriage. Though rarely printed in full on the ticket, but normally produced in a separate booklet, the passenger's attention should be drawn to their existence on the ticket. As we have already seen, such terms and conditions must be accepted by all parties before they can become incorporated into the contract and enforced in law.

Enforcement

The substance of Article 3 (2) deals with the effect of non-compliance with Article 3 (1). For carriers this can prove the most vital provision of the Convention. If, with the *consent* of the carrier, the passenger embarks without a proper passenger ticket having been delivered, or the ticket does not include the notice required by Article 3 (1)(*c*) the carrier shall not be entitled to avail itself of the limitation of liability provided for in Article 22. Article 22 will be considered in more detail at a later stage but, suffice it to say here, that there is a limit (relatively low by modern day standards) which could be of considerable value to the carrier.

We originally looked at the fundamental concept in the minds of the creators of the Warsaw Convention. We found that their aim was to provide a uniform system based on a fair balance between the opposing interests of the carrier on the one hand, and its passengers and consignors of cargo on the other, in the event of damage to persons and property during the carriage by air. The balance was achieved by giving the passenger/consignor the benefit (in almost all cases) of an assumed liability (in short, without proof of negligence) on behalf of the carrier in return for financial limits on that liability, thus protecting the carrier from unlimited liability claims. So, from the carrier's point of view, the assumption of automatic liability was balanced by limited liability. The benefit of these limits can, however, be lost if, for example, the carrier does not comply with the express provisions in Article 3.

As regards the rest of the contract the Convention states that the absence, irregularity or loss of the passenger ticket shall not affect the existence or validity of the contract of carriage which shall continue to be subject to the Convention.

Baggage check

Article 4 deals with documentation for baggage, referred to as a baggage check. This must likewise be delivered and may be incorporated in a passenger ticket delivered in accordance with Article 3 or, if separate, the baggage check must contain the same information, that is, in summary:

- places of departure and destination;
- agreed stopping place (if departure and destination in single territory); and
- Warsaw notice in respect of loss or damage to baggage.

Similarly the baggage check constitutes prima facie (but not conclusively) evidence of the conditions of carriage and of the registration of the baggage. The absence, irregularity or loss of the baggage check does not affect the existence or the validity of the contract of carriage which remains subject to the Convention. Again, however, non-compliance with the notice provision shall void the carrier's entitlement to limitation of liability in Article 22 (Article 4 (2)). A carrier must not take charge of baggage check having been delivered.

Significantly this provision relates to *registered* baggage as expressly stated in Article 4 (1). This is not defined but would clearly seem to be that baggage of which the carrier takes possession and carries normally, like cargo, in the hold of the aircraft out of the control of the passenger. By contrast the baggage which passengers keep and carry on board themselves, commonly known as hand luggage, is clearly not registered baggage. This is to some extent covered by Article 22 (3) (see p. 190).

Unregistered baggage has been something of a problem area in that whilst the Convention does not expressly impose a liability on the Carrier under Section 18, it does provide for a liability limit of 5000 Convention francs in respect of objects which the passenger takes charge himself (see Article 22 (3) and p. 190). It is therefore assured that in most cases the Convention does apply in respect of unregistered baggage. The passenger would rely on the ordinary law of tort, or possibly breach of contract. The likelihood of proving the carrier liable for loss or damage would be variable depending on the facts of the case. Without the benefit of the Convention the passenger would have to prove that the carrier caused the damage and was responsible, but on the other hand there would not be a statutory limit on the damages that could be claimed. As will be seen from our later examination of the provisions of liability, the Convention limits are relatively speaking very low and unlikely to cover anything of real value. In some circumstances, therefore, valuable items would be better taken as hand luggage where loss or damage caused by the carrier could result in a claim for a realistic amount on behalf of the passenger. This, of course, is dependent on a number of various factors. It may well be, for example, that the carrier's conditions of carriage contain provisions relating to unregistered baggage not covered by the Convention. This situation is a good example of where the Convention does not cover all situations and eventualities. There is still room for the prudent carrier to make contractual provisions to protect itself. This area will be further explored when looking at cargo and the specific problems relating to carriage thereof (see below).

Registered baggage will, of course, be within the Convention. The point of registration is, in effect, when the luggage is checked in by the passenger at the check-in desk and the baggage is weighed and taken by the airline. In practice the check-in staff produce and attach to the ticket a stub for each piece of checked-in or registered baggage. The baggage is then in the custody of the carrier and is subject to the Convention.

Air waybill

Similar provisions also relate to air waybills, the documentation for cargo. The situation is slightly more complicated, however, because of the different parties that may be involved. Article 8 states the air waybill must contain the usual information already discussed: the places of departure and destination; an agreed stopping place (where departure and destination are within the territory of a single High Contracting Party); and the Warsaw notice relating to the loss of or damage to cargo. The absence, irregularity or loss of this document does not affect the existence or validity of the contract of carriage which remains subject to the Convention. The usual provisions also apply in relation to contravention.

Article 9 states that if, with the consent of the carrier, cargo is loaded on board the aircraft without an air waybill having been made out, or if the air waybill does not contain the notice required by Article 8 (c) (notably the Warsaw notice), the carrier shall not be entitled to avail itself of the provisions of Article 22 (2), i.e. the limitation of liability is lost. Again there are two vital points:

(1) the documentation must be completed and delivered before embarkation; and
(2) the documentation must contain the specified information.

It must be stressed that a carrier could suffer the consequences of open-ended liability if it does not strictly adhere to these provisions.

The reader should review carefully the text of Section 3 on air waybills. Apart from the standard provisions stated above, similar to those of the passenger ticket and baggage check, there are a number of additional provisions necessitated by the particular circumstances posed by the carriage of cargo. In brief, some of these are referred to below.

The carriage of cargo usually involves three parties rather than two. Apart from the carrier, there is the person who is sending cargo, and the person at the other end to whom it is being sent. The shipper, or person who enters into the contract of carriage is referred to in the Convention as the consignor. It consigns the goods to the receiving party known as the consignee. Both the names of the consignor and consignee should appear on the air waybill. The consignor is, of course, the party contracting with the carrier, whereas the consignee is the party to whom the carrier must deliver the goods. The air waybill must, according to the Convention, be produced in various copies and each delivered as specified.

The carrier has the right to require that the consignor makes out and hands to it the air waybill; conversely, the consignor has the right to require the carrier to accept the document. The document must consist of (at least) three original parts. (An air waybill is often produced like an individual booklet with many duplicate pages to be distributed to a variety of persons as a record of the carriage.) As for the purposes of the Convention there must be three original copies of each distributed as particularised:

(1) The first part must be marked 'for the Carrier' and must be signed by the consignor. Clearly this is the part for the carrier evidencing for its purposes the contract of carriage.
(2) The second part must be marked 'for the Consignee'. It must be signed by

both the consignor and the carrier and must accompany the cargo. Clearly this is the part to be handed to the consignee with the cargo.

(3) The third part must be signed by the carrier and handed to the consignor after the cargo has been accepted. Clearly this is to confirm to the consignor the contract of carriage.

The carrier is to sign prior to the loading of the cargo on board the aircraft. For administrative convenience, the Convention allows the carrier's signature to be stamped and that of the consignor to be printed or stamped. If, at the request of the consignor, the carrier makes out the air waybill, it shall be deemed, subject to proof to the contrary, to have done so on behalf of the consignor. Where there is more than one package, the carrier has the right to require the consignor to make out separate air waybills.

This specific procedure (laid down in Articles 6 and 7) is carefully conceived to produce essential records for the various parties. It produces the evidence of the contract and its completion by communicating the acceptance of the cargo back via the carrier to the consignor. Unfortunately, however, these regulations are all too often flouted. The carriage of cargo involves many problems quite different from those posed by the carriage of passengers and their baggage. Difficult circumstances, involving for example animals, and urgent or emergency cargo are often dealt with by the cargo carrier in cargo aircraft and do not streamline well into a standardised administrative procedure. Half the office staff can end up in the cargo shed and the bits of paper may be dealt with later when the aircraft has taken off and peace returns. This, of course, is too late as far as the Convention requirements are concerned. The stringent requirements of the Convention specify in particular, not only the correct inclusion of certain information, but that certain documents be completed and delivered to certain persons within a time limit. Essentially this is *before* the aircraft takes off.

Any carrier of cargo should read and consider the original text of the Act with care. Reference to Article 9 will remind one of the consequences of non-compliance. The carrier will be open to unlimited liability if it allows the cargo to be *loaded* on board the aircraft without an air waybill, or if the air waybill does not include the Warsaw notice. It should be noted that the operative word is *loaded* and not take-off or departure of the aircraft.

Further provisions are also made in the Convention as to statements in the documentation. Article 10 makes the consignor responsible for the correctness of the particulars and statements relating to the cargo which it inserts in the air waybill. The consignor is also liable to indemnify the carrier against all damage suffered either by the carrier or any other person to whom the carrier is liable, by reason of the irregularity, incorrectness or incompleteness of the particulars or statements furnished by the consignor.

This is an important protection for the carrier. It clearly has to rely on the information about the cargo it receives from the consignor and there are, for instance, circumstances where the carrier might unknowingly carry illegal contraband such as weapons or drugs. Unless the carrier is aware of such defects in the particulars, the provision of Article 10 goes a long way to protect it from the false or fraudulent documentation produced by the consignor. In practice, the carriage of illegal or wrongful cargo is an ever-present problem for the cargo carrier.

Article 11 states, that, as with the passenger ticket and baggage check, the air waybill is prima facie evidence of the conclusion of the contract, of the receipt of the cargo and of the conditions of carriage. This means that in the first instance it is assumed to be so but can always be proved to be otherwise. In other words, prima facie evidence is not absolute proof.

As the carriage of cargo involves obtaining more information on the nature of the item to be carried than is necessary or relevant with human carriage or its baggage, provisions are also made regarding particulars of the cargo. They are broadly divided into two categories, 'weight, dimensions and packing' and 'quantity, volume and condition'.

Article 11 (2) provides that statements in the air waybill relating to the weight, dimensions and packing of the cargo (as well as those relating to the numbers of packages) are prima facie evidence of the facts stated. Those, however, relating to quantity, volume and condition of the cargo do *not* constitute evidence against the carrier except in so far as they have been (and are stated in the air waybill to have been) checked by the carrier in the presence of the consignor, or relate to the apparent condition of the cargo. The carrier is, therefore, largely protected provided that it has not checked and discovered otherwise in the presence of the consignor and so stated in the air waybill. Alternatively there is an exception relating to the apparent condition of the cargo. This can cause problems of interpretation but is clearly intended to cover a situation where, for example, the condition of the cargo is so visibly obvious that it is clear evidence as to its true condition.

Article 12 tries to deal with some of the practical problems arising out of the consignor's wish to alter arrangements for the cargo. It is given certain rights to make changes but not at the expense of the carrier or other consignors. The consignor is given the right (subject to its liability to carry out its obligations under the contract of carriage) to dispose of the cargo by withdrawing it at an aerodrome of departure or destination, or by stopping it in the course of its journey on any landing, or by calling for it to be delivered at the place of destination or in the course of the journey to a person other than the consignee named in the air waybill or by requiring it to be returned to the aerodrome of departure. The vital qualification is that the consignor must not exercise this right of disposition in such a way as to prejudice the carrier or other consignors, and it must repay any expenses occasioned by the exercise of this right.

This provision, in general terms, grants the consignor the right (subject to the qualifications stated) to recover the cargo from the carrier at an airport (being of departure, destination or stopping place). It cannot normally intervene if the cargo is in the course of its journey inbetween such places unless it is being taken to a person other than the consignee named in the waybill. This rather more drastic right is clearly to protect the consignor in circumstances where the carrier is presumably not carrying out the terms of the air waybill by failing to deliver the goods to the consignee therein named.

Article 12 continues to clarify these provisions. The carrier must inform the consignor forthwith if it is impossible to carry out its orders. If, however, the carrier obeys the orders, the carrier should ensure that the consignor produces that part of the air waybill delivered to it. Cargo should only be handed over by the carrier to the consignor on production by the latter of its part of the air waybill. For, the Convention states that if the carrier does not require the production of that document it will be liable (without prejudice to its right of recovery from the consignor) for any damage that may thereby be caused to any person who is lawfully in possession of that part of the air waybill.

Subsection (4) states that the right conferred on the consignor ceases at the moment when that of the Consignee begins, in accordance with Article 13 (see below). Nevertheless, if the consignee declines to accept the air waybill or the cargo, or if it cannot be communicated with, the consignor resumes its right of disposition.

Article 13 deals with the circumstances of normal delivery of the cargo. It sets out the procedure to be followed (other than where there has been a disposition of the cargo as set out in the previous Article 12).

(1) It is the duty of the carrier (unless otherwise agreed) to give notice to the consignee as soon as the cargo arrives.
(2) The consignee is entitled, on the arrival of the cargo at the place of destination, to require the carrier to:

(a) hand over the air waybill; and
(b) deliver to it the *cargo*.

(3) These rights of the consignee arise on its payment of the charges due and on its complying with the conditions of carriage set out in the air waybill.

Where the cargo has been lost or delayed Article 13 (3) provides that if

(1) the carrier admits the loss of the cargo; or
(2) if the cargo has not arrived at the expiration of 7 days after the date on which it ought to have arrived,

the consignee is entitled to put into force against the carrier rights which flow from the contract of carriage.

This seems to make good sense for if the cargo is lost or delayed the consignee can take action against the carrier on the basis of the contract of carriage. As will already have been seen in Chapter 25, under the law of contract the parties thereto can enforce the terms of the contract in a court of law. The consignee would not normally, of course, be a party to the contract of carriage entered into between the consignor and the carrier. Therefore, it could not normally enforce any term of the contract against the carrier. However, these rights can be granted, as here, by statute. This means that because of the express terms of the Convention, as incorporated here by the Carriage by Air Act 1961, the consignee can enforce the rights stated by the Act. These rights are, of course, limited to those expressly and specifically given by the Act.

Thus in Article 13, and indeed as we will see in Article 14, the consignee is given statutory rights which it would not normally enjoy under the law of contract. Article 14 states that the consignor *and the consignee* can each enforce all the rights given them by Articles 12 and 13 (each in its own name), whether it is acting in its own interest or in the interest of another, provided that it carries out the obligations imposed by the contract. For the avoidance of doubt, Article 15 states that Articles 12, 13 and 14 do not affect either the relations of the consignor or the consignee with each other, or the mutual relation of third parties whose rights are derived either from the consignor or from the consignee.

Two other general points are dealt with in Article 15. First, and important to note, it states that the provisions of Articles 12, 13 and 14 can only be varied by express provision in the air waybill. Significantly then it is possible to vary or even exclude these important provisions. It is essential, however, that in order to vary the statutory provisions, the terms must be expressly stated in the air waybill. If any variations do not conform with the provisions they will cease to be enforceable and the provisions of Articles 12, 13 and 14, as applicable, will stand.

This is an area where, despite the extensive and mostly mandatory provisions of the Convention, the carrier can intervene with its own contractual terms. It is always, therefore, greatly to the advantage of the carrier to review its potential liabilities and draw up additional terms of carriage which can apply within the terms of the Convention as well as those which are effective when the Convention is not applicable.

Finally, Article 15 states that, for the avoidance of doubt, nothing in the Convention prevents the issue of a negotiable air waybill.

Article 16 is the final article in Chapter 2, Section 3 which deals with air waybills. This article is primarily of protection to the carrier in order for it to properly comply with customs requirements in relation to due disclosure. The article provides that the consignor must furnish such information and attach to the air waybill such documents as are necessary to meet the formalities of 'customs, octroi, or police' before that cargo can be delivered to the consignee. As in similar provisions the Convention then goes on to provide for defects in such information.

The consignor is liable to the carrier for any damage occasioned by the absence, insufficiency, or irregularity of any such information or documents, unless the damage is due to the fault of the carrier or its servants or agents. Furthermore, the carrier is under no obligation to enquire into the correctness or sufficiency of such information or documents. This clarifies the carrier's position with more certainty where it could, in practice, be vulnerable.

In the carriage of cargo the carrier can find itself in many tricky situations.

One is, of course, where the cargo may be different from that which it was led to believe, and where the possession, carriage or importation of which would involve a breach of national or international laws or regulations. The obvious example is the disguised importation of prohibited drugs or weapons.

The carrier must be protected from the consequences of such deception on the part of the consignor or others. It may not be totally obvious where to draw the line which apportions all responsibility to the consignor. To avoid claims that the carrier ought to have known or guessed the true position, Article 16 makes it clear (that for these purposes) the carrier is under no obligation to make enquiries as to the validity of the information it has been given. Questioning to the point of actual proof could become an administrative impossibility: the carrier can merely accept information it receives from the consignor. At the other extreme, however, it does not mean that the carrier can or should accept obviously defective information, or even worse, knowingly hide behind information which it knows to be false.

Liability of the carrier

The next two Articles, namely 17 and 18, are the crux of the Convention. They specify the nature and extent of the carrier's liability. The essential terms are in Article 17 as applicable to passengers, and re-expressed in Article 18 as they relate to registered baggage and cargo.

Article 17 states

> The carrier is liable for damage sustained in the event of death or wounding of the passenger or any other bodily injury suffered by a passenger, if the accident which caused the damage so sustained took place on board the aircraft or in the course of any of the operations of embarking or disembarking.

The period of liability is *restricted in time* and *place*. The place must be:

(1) on an aircraft, or
(2) when embarking or disembarking from an aircraft.

The period of time must be that during which the subject of the damage, the passenger, is in the specified place, i.e. on the aircraft or embarking or disembarking onto or from it. Barring a few specific exceptions, the carrier is automatically liable in the circumstances stated by the Convention. Unlike the normal tort of negligence the claimant would not have to prove that the carrier owed it a duty of care, that that duty had been breached, and that the breach directly caused the damage. We therefore return to the essence of the Convention: a balance of benefits allowing for overall uniformity and fairness in general terms – the carrier accepts automatic liability but there are limitations on the monetary value of that liability. The passenger or consignor of cargo who has suffered damage can claim more easily (normally) without proof of fault, and therefore (normally) without the need to go to court, but on the other hand must accept the Convention limits on damages awarded and cannot 'sue for the sky' with open-ended limitless claims.

Let us now consider more precisely the meaning of Article 17. Along with Article 1, Article 17 is a vital provision of the Convention, but the seemingly precise terminology can require further clarification in use. As with Article 1, we shall, therefore, consider certain of the essential words or phrases.

Accident

This word appears in the middle of the Article and is in fact the prime word, for without the occurrence of an accident, the trigger event, there can be no damage covered by the terms of the Convention. There is no further definition of 'accident' given in the Convention or the Act. It is clear that there must have taken place a specific and positive occurrence that caused the damage sustained. A useful definition by way of reference is that given in Section 75 (4) of the Civil Aviation Act 1982. An 'accident' is construed as 'including any fortuitous or unexpected event by which the safety of an aircraft or any person is threatened'.

Past decisions and commonsense establish general guidelines but there are still areas of doubt and debate. The best test is perhaps that of normality. Normal conditions could not easily be construed as an accident, whereas the occurrence of something unusual or unexpected could. For instance, an individual's harm or discomfort during normal flight conditions without incidence are unlikely to be claims within the Convention. Attempted claims have been made, but unsuccessfully, by passengers where damage could not really be said to have been caused by any particular or unexpected occurrence during the flight, and was not therefore construed as an 'accident'. Often these types of claim involve injury or death caused by inherent problems with the victim's health. For example, where an asthmatic with a weak heart suddenly has a heart attack and dies, or a passenger with a hernia suffers discomfort for no particular reason during a normal flight. In these cases there appears to be no 'accident' as such which caused the harm.

By contrast, sudden turbulence causing anything from jolts, falls, burns from spilled hot drinks, and loss of hearing and balance and so on can be construed as the result of an accident. Similarly an unusually heavy landing may be an 'accident'. It all depends on the nature of the occurrence – to what extent was it unusual or unexpected? It is often a matter of degree in which each individual case has to be examined on its own facts.

It must also be remembered that no element of fault on the part of the carrier has to be proven. Thus, even where, for example, objects fall from an overhead locker and injure a passenger below, the carrier would normally be liable. The accident would be the sudden falling of objects and even where, say, that locker had been left open by another passenger, the carrier would in the first place be liable to the injured party within the terms of the Convention.

Passenger

We have established that an 'accident' must occur but to whom must this cause damage? The answer is expressly stated as a 'passenger'. There has been some doubt and subsequent debate about what falls within this definition. 'Passenger' is not defined, though a commonsense construction would suggest the use of that word implies the non-inclusion of crew. Article 3, dealing with tickets, also refers to 'passenger'. However, Article 1 relating to the applicability of the Convention refers to 'persons'. The net result is that in most cases working crew are not within the Convention for these purposes and in any case a different legal relationship exists by virtue of their position as an employee of the carrier. This may not stop a non-working crew or employee being construed as a passenger.

Bodily injury

Clearly there must be damage sustained as a result of the 'accident' for any claim to exist. The nature of the damage is defined in three categories:

- 'Death'
- 'Wounding'
- 'Any other bodily injury'

Physical injury is clearly within the definition but again there are grey areas in practice. The main problem area has been that of mental suffering or distress, whilst injury is specifically qualified by the word 'bodily'. There has been an increasing tendency to include mental injury. This reflects the developments in recent years which have extended the area of claims from mental suffering, particularly in relation to negligence. As seen in the discussions on the tort of negligence, there has been a continual widening of the net by the courts. These developments have probably been most extensive in the USA but have occurred to some extent in other jurisdictions.

'Any of the operations of embarking or disembarking'

It is these words that have caused the most controversy and have resulted in varying constructions being applied at different times and places. The most narrow interpretation would be literally just stepping on or off the aircraft, perhaps mounting or descending the steps (which would have been the method of access at the time the Convention was written). What is significant is the phrase 'any of the operations' which tends to widen simply embarking or disembarking and has provided the potential for extending the application in modern times.

Of course, the actual physical process has changed, becoming more complicated and sophisticated. In many airports feeder arms, often known as air bridges, link airport buildings with aircraft and there is not a break between the building and reaching and entering the aircraft and vice versa on disembarkation. Often long walkways lead to the air bridges, all of which extend from the departure channel from which, once entered, there is no point of return. In a sense these are all part of the physical process of boarding an aircraft. Likewise, on disembarkation, are the walks to and progression through luggage reclaim, customs, and immigration. Where does disembarkation end and embarkation start?

In one sense the process of embarkation starts at the check-in counter, or perhaps even at arrival at the airport, the official place of departure. Where to draw the line has proved a difficult problem. Where it is drawn has significant consequences. For, in the event of an accident, whether or not the Convention is applicable could affect:

● whether or not the carrier is liable; and
● whether or not there is a limit on the carrier's potential liability

The advantages and disadvantages to each party, as to whether the Convention applies or not has been repeated before but can never be overstated.

In summary, if the Convention applies the passenger has the advantages of a far easier burden of proof (negligence does not have to be proven), but the disadvantage is that the carrier's liability is limited to a relatively low amount in most jurisdictions. If a claim for negligence would be unlikely to succeed, a passenger would perhaps prefer to be under the umbrella of the Convention, knowing that there would be at least a better chance of receiving something. However, in the event of clear negligence on behalf of the carrier, and a potentially large claim, it would be to the disadvantage of the passenger (but clearly to the advantage of the carrier) for the claim to be within the Convention. Certainly the limitation on liability is a valuable asset to airlines and has become increasingly so as the gap between the fixed Convention limits and the ever-mounting civil damages awards by courts in many jurisdictions has increased.

Although there was some simplicity in the words 'embarkation' and 'disembarkation' as originally construed, the courts in certain jurisdictions, particularly in the USA, have been tempted to stretch the meaning to limits probably well outside the minds of the drafters of the Convention. For example, in the USA the Day–Evangelinos test is often applied (after the Day–Evangelinos cases). These two cases, one brought by Day in New York, and the other by Evangelinos in Pittsburgh, followed terrorist attacks on passengers in an airport boarding lounge at Hellenikon Airport Athens, in 1973. The question was were those passengers harmed whilst in the process of 'any of the operations of embarking or disembarking'. Eventually it was held that this was within the Convention in that the passengers were in the process of embarkation. There was, however, considerable debate and some dissension with the original decisions in each case reaching a contrary conclusion. After several appeals, a consensus resulted and final judgments agreed that the Convention was applicable.

It is, of course, quite possible and, indeed, probable to argue that the original drafters had no intention whatsoever of extending the carrier's liability to such places and occupations as sitting in an airport lounge. There is also an argument, however, that the drafters would have intended the Convention to be updated so that it continued its development as a living structure relevant to current times. Indeed, a rigid and increasingly out-of-date and out-of-touch system would risk its very survival. These issues will be more apparent when reviewing the actual monetary limits set under Article 22.

In conclusion, it has to be said that there is no definite answer applicable in any time or jurisdiction as to where and when the operations of embarkation and disembarkation begin and end and thus when exactly the Convention is applicable. Clearly certain jurisdictions, notably the USA, will continue to widen the construction in an attempt to make the provisions reflect the developments in civil liability generally. By contrast, in the jurisdictions of many other High Contracting Parties the interpretation has either not been recently tested or has a more restricted construction. As time goes on differences increase and the original purpose of the Warsaw System to provide a uniform system of regulation is under attack.

Article 18 similarly deals with the carrier's liability in relation to registered baggage and cargo. Article 18 (1) states that

> The carrier is *liable* for damage sustained in the event of the *destruction or loss of*, *or damage to*, any *registered baggage* or *any cargo*, if the occurrence which caused the damage so sustained took place *during the carriage by air*.
>
> (emphasis added)

'Liable'

Again it must be stressed that this implies strict liability on behalf of the carrier: so long as the damage occurs, as defined, the carrier is liable irrespective of fault.

'Damage'

The type of damage to be suffered is defined in three parts:

(1) 'Destruction', or
(2) 'Loss', or
(3) 'Damage'.

This appears fairly wide and includes damage in general terms as well as destruction or loss. As we have already seen (p. 162) Section 4A of the Act provides that damage is to be construed as to include *loss* of part of the baggage or cargo in question.

'Carriage by air'

This clearly needs further investigation and is defined in Article 18 (2), as comprising:

> the period during which the baggage or cargo is in charge of the carrier, whether in an aerodrome or on board an aircraft, or, in the case of a landing outside an aerodrome, in any place whatsoever.

This extends the period and places of the carrier's liability quite considerably from the act of embarking, say loading, and disembarking, say, unloading, used in the case of passengers. The important qualification is that the baggage or cargo must be in the charge of the carrier. There are three places specified during the relevant period of the carriage by air:

(1) An aerodrome, or
(2) on board an aircraft, or
(3) in the event of a landing outside an aerodrome in any place whatsoever.

Clearly it would be expected that cargo or registered baggage would be covered by the Convention whilst on the aircraft.

Extending the Convention to include an aerodrome will sensibly include loading and unloading during the period in which, inevitably with non-human objects, the goods will have to be in the control of the carrier before being delivered to a third party. The potential period of cover, or in other words the time during which the carrier remains liable under the Convention, is thereby extended considerably from that applicable in the case of passengers. It would seem that even where the cargo has cleared customs and waits for days or more under the control of the carrier whilst on airport premises it is within the Convention. Thus, a carrier may be liable for damage caused, even by a third party, to such goods whilst in a warehouse.

The third category of liability actually extends outside the aerodrome to 'any place whatsoever' but is strictly limited to cases where there has been a 'landing outside an aerodrome'. In other words where, for instance, there was a forced landing outside an airport the carrier is in charge of the cargo or baggage. The carrier can then find itself responsible for damage until such time as it ceases to be in charge of the baggage or cargo. It must also, of course, ensure that it complies with all the provisions relating to cargo and baggage and any additional contractual provisions.

An English case, *Swiss Bank Corporation* v. *Brinks MAT Limited* (1985) illustrates the workings of Article 17 in relation to cargo. This case concerns the theft of bank notes from a warehouse at London's Heathrow Airport.

The consignment of bank notes was delivered to the warehouse of the carrier, KLM, at the airport. Two of the three consignments were checked and weighed under the carrier's supervision. The third consignment was about to be checked when robbers burst in and seized the lot. It was held by the court that the carrier was liable for the loss of the first two consignments but not the third. The reasoning being that the consignments did not come into the charge of the carrier when the van containing the cargo, locked and guarded by the consignor's agents, entered the warehouse. At that stage the carrier's staff had no control over the cargo or its guardians and certainly no right of access into the van. The point at which the cargo came into the control of the airline was when it had been unloaded, checked and weighed.

Article 18 (3) deals with a situation where carriage by land, sea or river may become part of the overall carriage of the goods outside the aerodrome. This section expressly states that the carriage by air does not include such carriage when performed *outside* the aerodrome. As we have seen, there are likely to be few instances where the carrier is liable outside the aircraft and the aerodrome, but where it is prima facie liable 'in any place whatsoever' it does not include carriage by land, sea or river outside the aerodrome (though normally road transport within an aerodrome for goods in charge of the carrier would be within Article 18). It should be noted that the reference to an aerodrome is likely, if certain foreign decisions are followed, to include all terminals and other buildings within the aerodrome perimeters.

Finally, Article 18 (3) also states that if such carriage does take place, in other words by land, carriage or sea outside an aerodrome, for the purpose of loading, delivery or transhipment any damage is presumed, subject to proof to the contrary, to have been the result of an event which took place during the carriage by air. This helps clarify a position of obvious doubt. All too often cargo can be damaged in circumstances where it is difficult, if not impossible, to prove at what point the damage occurred. The Convention here tries to give certainty where there could be doubt and brings such situations within the scope of Article 18 unless it can be proved to the contrary that damage occurred at a time excluded by Article 18 (3). In other words during carriage by land, sea or river performed outside an aerodrome.

Delay

Article 19 deals with the all too common problem of delay. It expressly states that the carrier is liable for damage occasioned by delay in the carriage by air of passengers, baggage or cargo. Interestingly it does not specifically refer to registered baggage as in other places and it is, therefore, arguable that all baggage whether registered or not, i.e. whether in the control of the carrier or the passenger, is covered by this provision. This is a seemingly absolute and useful provision for the benefit of the passenger and the consignor of cargo.

However, in practice this is rarely invoked for in reality delays, though frequent, are normally short and no quantifiable damage is suffered.

It is clearly not worth a passenger claiming damages for an hour or so's delay. Despite the inconvenience and irritation caused, there will normally be no real harm suffered. Airlines usually have a policy of placating passengers in such circumstances by offering, for example, free vouchers for food and other benefits as a gesture of goodwill. Claims for passenger delay are, therefore, rare and are not normally a problem for the carrier if handled with consideration and sensitivity at the time. Exceptional circumstances may, of course, occur where a delay becomes significant causing a passenger loss such as a business traveller failing to reach an important meeting, contract completion or tender deadline. Such unusual circumstances may give rise to considerable claims.

Delay is normally more relevant and problematic in terms of the carriage of cargo, particularly for cargo operators. The carriage of time-sensitive goods, such as perishable stock or animals, can be full of difficulties if delay is incurred. Dedicated cargo operators, in particular, need to review their legal position in relation to potentially problematic cargo. For instance, day-old chicks have often been exported by air to African countries. The timing of the flight is vital in that if they are carried within so many hours of hatching the chicks travel well and require no feeding or attention. Delay beyond this period can, however, be fatal so that a delayed take-off of several hours can cause the death of the chicks for which the carrier may well be liable.

Over-booking

Before leaving the problem of delay, reference should be made to similar problems that may arise in practice. It is well known that airlines in effect carry out a policy of over-booking where demand allows. Airlines claim that this is a commercially justifiable practice, in particular to counteract the 'no shows'. These are passengers with bookings for a flight who then fail to turn up. Commercially the airline will lose the potential sale of that seat, at least in the circumstances where the ticket can be used on another occasion. With modern computerised booking of seat allocation systems it is possible to predict, with a good degree of success, the amount of 'no shows' on a particular flight and counteract this with 'double bookings'. Of course, this assumes the demand is there in the first place but on appropriate flights airlines have become experienced in 'playing the odds', and overbooking only up to the 'no-show' limit. Of course, they have to make a prediction that in the event may not be correct: at worst 'no shows' will be less than the over-bookings — net result, *bumping*.

Excess of booked passengers over capacity will mean some passengers will be unable to go on the flight for which they were booked. Clearly the carrier is not in a good position here. Legally it is usually in the wrong. Quite apart from any Convention considerations, the carrier has prima facie breached its contract to carry the passenger on a particular flight. In theory there are a number of legal considerations; in practice, whilst hopefully the situation rarely occurs, it is normally dealt with by an appropriate placation of individual passengers. This will obviously have to involve re-booking on the earliest alternative flight and often includes some form of monetary or other compensation. One known way of dealing with the problem is, rather than the airline selecting certain passengers to 'bump', all passengers are offered an inducement, normally a cash payment, to go on the next available flight. This discreet method prejudices no particular passengers and cleverly encourages the passengers to voluntarily select themselves for a later flight on the basis that they are happy to receive the inducement. Subsequently claims are rarely made for 'bumping' and the problem is dealt with on a practical, individual basis without the carrier losing too much goodwill.

The legal situation is not particularly clear or helpful and it is best for all parties that a practical solution as indicated above is found. A claimant, a 'bumped' passenger, may in theory have several potential claims (though the recent European Community regulations discussed on p. 186 should now resolve the situation).

The first possibility must be that of breach of contract. Theoretically the passenger is likely to have a good case in terms of proving breach of contract, but the remedy for breach is damages and will be based on a monetary assessment of the harm caused. This is where the claimant's case is likely to be weak or, at worst, a waste of time. What can the average passenger claim in circumstances where he is put on a later flight? Yes, there has been inconvenience, irritation and maybe arrangements with third parties have had to be altered, but there is rarely any significant assessable loss in such circumstances. So, unless there were unusual facts, such as proof of considerable monetary loss caused by delay, such a claim and any court proceedings are likely to be a waste of time and money on the part of the passenger. The airlines are, of course, well aware of this and are therefore encouraged to continue the practice.

There may be other forms of action to consider, for instance under the Trades Description or other consumer protection legislation or for misrepresentation, but again, in reality, it is not likely to be worthwhile in normal circumstances (see *British Airways Board* v. *Taylor* [1976] 1 ALL ER 65). Whilst it would seem in legal terms that the obvious claim is for breach of contract, there has been some debate as to how the practice of 'bumping' relates to Article 19 of the Convention. The question is, can 'bumping' be classified as a form of delay? Unfortunately, the answer is far from clear.

It would seem that in a precise interpretation delay does not cover 'bumping'. Delay surely presumes that first, there was a delay, in that a particular flight could not take off when scheduled but eventually departed at a later time, and second, the passengers therefore suffered delay but eventually departed *on that particular flight*. By contrast, in a 'bumping' situation there is no assumption of delay of the relevant flight, indeed it would presumably depart on time and, if not, it is not particularly relevant: what is relevant is that the 'bumped' passenger did not board that particular flight but had to go on another one. Despite this logical argument, attempts have been made to bring 'bumping' within the scope of Article 19 (and therefore classify it as a delay) but there has been no clear English authority on this point and the foreign trend is not to include 'bumping' within the scope of the Article.

It should be noted that if, in the unlikely event that 'bumping' is construed as being a form of delay and therefore subject to the provisions of Article 19, there is little the carrier can do to exclude or limit its liability thereunder. If, however, as is more likely, 'bumping' is not within Article 19 the carrier may be able to exclude or limit its potential liability by introducing some protective wording into its conditions of carriage. For instance, the carrier may provide that the purchase of a ticket does not guarantee a seat on a particular flight. It should also be noted that whilst there may be a variety of civil liabilities involved with 'bumping', it is also possible there may be criminal consequences under English law, though this has not been properly tested.

European Community regulations

Fortunately, much of the theoretical debate should now be replaced by the recent European Community regulations which establish common rules for a system of denied-boarding compensation in scheduled air transport. These regulations came into force on 8 April 1991. The rules apply where passengers are denied access to an over-booked scheduled flight for which they have a valid ticket and a confirmed reservation departing from an airport located in the territory of a member state to which the Treaty applies (irrespective of the state where the air carrier is established, the nationality of the passenger and the point of destination). The regulations clarify by further definition many of those terms.

In brief, the regulations provide that carriers to which they apply must make rules for over-booking which are to be filed with the appropriate authorities and made available to the public. If boarding is denied a passenger is to have the choice of:

— reimbursement without penalty of the cost of the ticket for the part of the journey not made, or
— re-routing to his final destination at the earliest opportunity, or
— re-routing at a later date at the passenger's convenience.

The rules provide for a minimum amount of compensation payable to the passenger. A carrier is not obliged to pay compensation in cases where the passenger is travelling free of charge or at reduced fares not available directly or indirectly to the public.

The rules in full are set out in the EEC Council Regulation (No. 295/91 of 4 February 1991) as reproduced by the CAA in its *Official Record*, Series 2, Air transport licensing notices, published on 9 April 1991. The CAA therein provides that carriers who are subject to the regulation should submit a copy of their rules (and any subsequent changes) to the following:

(1) The Civil Aviation Authority
    Room T522, CAA House, 45–59 Kingsway, London WC2B 6TE
(2) Air Transport Division, DG VII
    Commission of the European Communities, Rue de la Loi 200, 1049 Brussels, Belgium
(3) Department of Transport
    1A3 Directorate, 2 Marsham Street, London SW1P 3EB

Carrier defences

Articles 20 and 21 may come as something of a relief to the carrier who may feel bound without escape by the previous provisions. We should remember, however, that though the carrier's liability is wide there are limitations on that liability. The carrier is also offered two possible defences.

'All necessary measures'

Article 20 provides an exclusion to the benefit of the carrier or its agent. It states that the carrier is not liable if it proves that it, its servants or agents have taken all necessary measures to avoid the damage or that it was impossible for it or them to take such measures. This is in a sense the one real 'get-out' clause in the Convention, so far as the carrier is concerned. There are several points to bear in mind. First, the words 'if he proves' make it plain that the onus of proof is on the carrier. In other words the contrary will be assumed unless it can actually prove that all necessary measures were taken. The words 'all necessary measures' are broad if not rather vague. The word 'all' suggests a rather high standard of proof is required. It is of course difficult to prove the absolute: to show that absolutely everything that could possibly be done was done to avoid the damage. Where is the line to be drawn? It is in practical terms an infinite standard. The qualification 'or that it was impossible for him or them to take such measures' is of some help. Though, again what is actually impossible is perhaps a rather imprecise standard.

In conclusion, the exclusion of liability in Article 20 is not often of help to the carrier, though it may frequently try to argue the point which may also be of some use in a bargaining situation where a settlement is reached between the parties without resorting to legal action. It should also be noted that, as we shall see later when considering the US position, this provision is exempted (in respect of passengers) in US flights by virtue of the Montreal agreement of 1966.

'Contributory negligence'

Article 21 deals with contributory negligence. It states that if the carrier proves that the damage was caused by or contributed to by the negligence of the injured person the court may, in accordance with the provisions of its own law, exonerate the carrier wholly or partly from liability. This is in practice a provision helpful to the carrier. Under English law there is an established common law defence of contributory negligence. This allows the defendant who is being sued for negligence to claim that all or part of the damage was caused by the plaintiff. It therefore tries to apportion fairly the blame, particularly where the fault may be shared, even if not equally, between the parties.

Potentially this sort of situation can easily arise: for instance where a passenger is hurt during turbulence having disobeyed a warning and instructions to remain seated with the seat-belt fastened, or where a passenger is burnt in a fire in the lavatory which was caused by the passenger smoking a cigarette which is specifically prohibited there. In such cases the passenger has clearly contributed in part, at least, to the harm caused to him. To the extent that the passenger has so contributed to the damage, the value of the claim against the carrier would be proportionately reduced. This will be in accordance with the law of the jurisdiction of the court hearing the case and may, therefore, exonerate the carrier wholly or partly from liability.

Disability limits

Article 22 contains a number of provisions and is a key part of the Convention for it specifies the actual limitations on the carrier's liability in financial terms. Article 22 (1) deals with the carriage of persons; Article 22 (2) deals with registered baggage and cargo, and Article 22 (3) deals with baggage of which the passenger takes care.

Passengers

Article 22 (1) states that

> In the carriage of persons the liability of the carrier for each passenger is limited to the sum of 250,000 francs. Where, in accordance with the law of the courts seized of the case, damages may be awarded in the form of periodic payments the equivalent capital value of the said payments shall not exceed 250,000 francs. Nevertheless, by special contract, the carrier and the passenger may agree to a higher limit of liability.

The monetary currency referred to is Convention Francs and the problems associated with this will be discussed below.

First, it should be noted that the limit is an overall limit applicable to each individual passenger. Second, damages can be awarded as a series of periodic payments and third, a higher (but not lower) limit of liability may be agreed by special contract. There is in fact a higher limit imposed in the UK by the CAA. Known as Condition H, the provision is imposed as a standard condition on route licences granted by the Authority. It is, therefore, likely to apply on all international routes operated by British airline operators. Condition H is defined in the CAA's *Official Record*, Air Transport Licensing, Series 1 (as at November 1990 'Standard Conditions for Air Transport

Licences', p. 34):

> The licence holder shall enter into a special contract with every passenger to be carried under this licence on or after 1st April 1981, or with a person acting on behalf of such a passenger, for the increase to not less than the sterling equivalent of 100,000 Special Drawing Rights, exclusive of costs, of the limit of the carrier's liability under Article 17 of the Warsaw Convention of 1929 and under Article 17 of that Convention as amended at The Hague in 1955.

The Convention Francs are converted into sterling at a rate designated by the government which is regularly updated by statutory instrument. The Carriage by Air (sterling equivalents) Order 1986 (SI 1986 No. 1778) specifies that

> the amounts shown in column 2 of the following Table are hereby specified as amounts to be taken for the purposes of Article 22 in the First Schedule to the Carriage by Air Act 1961 and of that Article as applied by the Carriage by Air Acts (Application and Provisions) Order 1967 as equivalent to the sums respectively expressed in francs in the same line in column 1 of that Table:

TABLE

Column 1 Amount in Francs	Column 2 Sterling Equivalent
	£
250	13–63
5,000	272–67
125,000	6,816–68
250,000	13,633–40

This Order is applicable as at 1 September 1991 but liable to be updated at any time and checks should always be made as to the current position.

Registered baggage and cargo

Article 22 (2) specifies that the liability of the carrier in respect of registered baggage and cargo shall be limited to 250 Convention Francs per kilogramme. This is to be so unless the passenger or consignor, as the case may be, has made, 'at the same time when the package was handed over to the carrier, a special declaration of interest in delivery at destination and has paid a supplementary sum if the case so requires'.

Again a standard limitation is imposed but with the option of the parties making a special agreement. In the case of goods it is appropriately different from that of passengers. The consignor or passenger has the right (so long as it is expressed before the goods are handed over), in effect, to declare a specific value for the goods. For this the passenger or consignor is required to pay, if necessary, for what would, in effect, be the cost of insurance. If this option is taken the carrier will then be liable up to the declared sum so long as that sum is not greater than the passenger's or consignor's actual interest in delivery at the destination. If, however, the carrier can prove that the declared sum is greater than that actual sum it will not be liable for the higher sum. The evaluation of the passenger's or consignor's interest in the delivery and the limiting in effect of the compensation to actual rather than declared values are normal principles of insurance. By taking the option of a special declaration of interest, the consignor or passenger is, in effect, taking out an insurance on the actual value of its interest in the delivery of the goods, be they cargo or registered baggage.

This special declaration is frequently and prudently used in relation to consignments of cargo. It is rarely used, however, in terms of registered baggage and it might be interesting to test the response of check-in staff if a passenger requested the making of a special declaration of interest in respect of baggage as it was handed over to them to be checked in.

Hand luggage

Article 22 (3) deals not with all hand luggage but with 'objects of which the passenger takes charge himself'. In respect of these items the liability of the carrier is limited to 5,000 Convention Francs per passenger. It should be noted that these 'objects' are not necessarily the balance of all items which the passenger has not checked in to become registered luggage. It is merely those items of which the passenger takes charge. What then happens if the passenger hands over a piece of hand luggage, be it an overcoat or a valuable painting, that is then taken by the cabin crew and hung or stored in the wardrobe space at the front of the aircraft? It seems that there has been created, by virtue of the wording of this provision, a fourth category of goods that travels in some capacity or another on the flight. We should consider therefore the following categories:

(1) The separate category of unaccompanied goods which constitutes cargo and is sent by a consignor for carriage independently of any passenger.
(2) Registered baggage, which is checked in by the accompanying passenger, taken by the carrier and normally carried in the hold of the aircraft.
(3) Unregistered baggage or goods which the passenger keeps in its own possession; essentially hand luggage which the passenger carries onto the aircraft.
(4) A further division of category 3, comprising those items in the possession of the passenger that are handed over to airline employees on the aircraft in contrast to those of which the passenger keeps possession.

In practice there are clearly instances where the distinctions are somewhat blurred, for example, the airline removing hand luggage to another part of the aircraft, perhaps the hold. In the latter case there is a query as to whether, by virtue of it being put into the hold, it then becomes registered baggage. The most likely interpretation is that once the carrier takes total charge of the goods and they are placed in the hold they become registered cargo, in which case the carrier must issue the proper documentation, in accordance with Article 3. However, goods are unlikely to be consigned as cargo simply because they were put for convenience in another part of the passenger cabin, particularly where the passenger has access.

Damages

The remaining subsections of Article 22, namely 22 (4) and 22 (5) deal with the amount of damages awarded under the Convention. The monetary value of damages and its calculation has caused many problems and today threatens the long-term future of the Warsaw System.

The initial problem was finding a unit of valuation that could be easily converted into the different currencies of the various High Contracting Parties. The unit chosen was the Convention Franc. At the time of the original unamended Warsaw Convention the French Franc was in fact used as the basis of an appropriate unit of calculation and Article 22 (4) of the unamended Convention refers expressly to the French Franc. Having found a suitable base unit, the founders of the Warsaw Convention thought the problem was solved. The mechanism for converting the Convention Franc into individual currencies lay in the fact that the Franc was pegged to the price of gold. From this countries could convert the Convention Francs through the price of gold to their national currency.

The Convention Franc has also been known as the Poincare Franc, so-named after a French administrator of the time, under whom the French Franc was stabilised in terms of gold. This meant that conversion into other currencies was relatively straightforward, and subject to the stabilising effect of the gold standard. With some forward perception the Convention founders also defined the value of the Convention Franc in relation to gold to avoid any doubt and, presumably, to protect against a position where the French Franc ceased to be pegged to gold. In the event, this, of course, happened, and with the decreasing importance of the gold standard over the ensuing years the method of conversion has become increasingly difficult.

Article 22 (5) of the Convention was accordingly amended at The Hague to read as follows:

> The sums mentioned in francs in this Article shall be deemed to refer to a currency unit consisting of sixty-five and a half milligrammes of gold of millesimal fineness nine hundred. These sums may be converted into national currencies in round figures. Conversion of the sums into national currencies other than gold shall, in case of judicial proceedings, be made according to the gold value of such currencies at the date of the judgement.

Thus conversion into national currencies via the gold standard seemed precise and secure and well in tune with the intention of the original Convention to provide a stable and uniform system.

Unfortunately, this has not worked as the original founders intended. Due to changes over the years in the different ways of stabilising and converting currencies relative to each other, the original method of conversion has basically become defunct. Some states have tried to adopt ways round this problem, whereas others, such as the USA, have used it as justification for attacking the limits themselves. As will be seen later, this has been the one fundamental flaw in the Warsaw System.

The remainder of Chapter 3 of the Convention deals with various miscellaneous provisions in relation to liability.

Exclusions on liability

Following on from Article 22 which specifies the actual limitations of liability, Article 23 deals with exclusions. It was obviously vital if the central ideals of the Convention were to apply with any uniformity that the carrier be prohibited from lowering the limits of its liability by contractual agreement. Article 23, therefore, states definitively that 'Any provision tending to relieve the carrier of liability or to fix a lower limit other than that which is laid down in this Convention shall be null and void.' For further clarification the Article provides that the nullity of any such provision shall not affect the remainder of the contract of carriage which shall remain subject to the Convention.

There is an important exception to protect the carrier against inherent problems with cargo. The provision in Article 23 (1) shall not apply to provisions governing loss or damage resulting from the inherent defect, quality or vice of the cargo carried. In other words the carrier is able to agree contractually lower limits than those laid down by the Convention where loss or damage results from inherent defects in the cargo.

Article 24 clarifies two further legal points. First, any action for damages arising out of Articles 18 and 19 (i.e. relating to registered baggage, cargo or delay) can only be brought subject to the conditions and limits set out in the Convention. Second, this is also to apply in respect of Article 17 (liability for passengers) but without prejudice to any questions as to who are the persons who have the right to bring suit and what are their respective rights.

Qualifications to liability

Articles 25 and 25A (the latter was inserted by The Hague Protocol), are important qualifications to the liability of the carrier. Article 25 specifies circumstances in which the carrier can not avail itself of limitations of liability (as specified in Article 22). These limits are not to apply, and therefore the carrier has no limit on its potential liability, where it is proved that the damage resulted from an act or omission of the carrier, its servants or agents done with either:

(1) intent to cause damage, or
(2) recklessly and with knowledge that damage will probably result.

Where the act or omission is that of a servant or agent of the carrier, it must also be proved that they were acting within the scope of their employment.

It is perhaps reasonable that there should be such an exclusion of the carrier's limitation of liability where it or its representative deliberately cause harm to a passenger, cargo or registered baggage, thus in effect abusing the protection afforded by the Convention. Article 25 tries to express the principle but the terminology again leaves the interpreter less than sure where the line is drawn.

The first condition, 'intent to cause damage', is clearer, though in practice proving real intent may be difficult and this is perhaps where the second condition is imposed as an alternative. 'Recklessly' falls short of deliberate intent to cause harm but implies a blatant and perhaps negligent disregard for a likely outcome. It is significant that there was some alteration in the wording of the original Warsaw Article 25 which referred to 'wilful misconduct', and was then substituted by the present terminology by the Hague Protocol. Recklessness still presents problems of interpretation as is well illustrated by the English case of *Goldman* v. *Thai Airways International Limited* (1983) CA. The decision of the High Court at first instance was reversed by the Court of Appeal in 1983.

The facts of the case were thus. The plaintiff, a Mr Goldman, was a passenger injured on an international flight with Thai Airways International Limited, the defendant. The plaintiff suffered a serious back injury when he was thrown from his seat as a result of severe clear air turbulence. The role of a pilot in relation to the turbulence was examined and certain pertinent factors became apparent. The pilot was aware that moderate clear air turbulence was likely and failed to take the appropriate action of switching on the cabin sign requiring that seat-belts be fastened, thus being in default of the instructions in the airline's manual. He was, therefore, clearly in breach of his duty. The question was how could he, or should he have foreseen the likely consequences.

The court at first instance held that in failing to illuminate the seat-belt sign the pilot had acted recklessly and with knowledge that damage would probably occur. The carrier, therefore, could not avail itself of the limits of liability specified in Article 22 and accordingly the plaintiff was awarded damages in excess of those limits. The defendant airline appealed against the decision which was, in the event, reversed by the Court of Appeal. The criteria of Article 25 were re-examined. The key question was, even if the omission was 'reckless' was it also done 'with the knowledge that damage would probably result'. Article 25 makes it quite plain, by the use of the word 'and' that both parts of the criterion, i.e. recklessness and knowledge, must be satisfied. It is not sufficient merely to be reckless. Could it, in this case, therefore, be proved that the pilot had knowledge that such damage would *probably* result?

It was held that because it was not proved that the pilot had such knowledge, the original judgment was set aside in so far as it related to excess damages. The result being that the plaintiff did not manage to avail himself of Article 22 in that the specific criteria against the carrier were not proved, thus the usual limits of Article 22 applied.

It has been said that as a result of this decision it will be almost impossible for a plaintiff in the English courts to discharge the burden of proof necessary to show that the carrier acted not only recklessly but 'with knowledge' that damage would probably result. The position may not be as extreme as that but it will certainly be difficult to prove the appropriate knowledge within this jurisdiction.

The *Goldman* v. *Thai Airways* case has had a significant impact in that many carriers now pay more attention to the effects of turbulence. For instance, in recent years it has become common practice to recommend that seat-belts remain fastened while a passenger is seated, and often notices to that effect are placed on the back of the seat in front.

Article 25A clarifies the position of 'servants' or 'agents' of the carrier covering, in particular, that of employees. In the original Warsaw Convention of 1929 there was, under Article 20 (2), a provision relieving the carrier of liability in respect of goods and baggage if it proved that the damage was occasioned by 'an error in piloting, in the handling of the aircraft, or in navigation and that, in all other respects, he and his agents have taken all necessary measures to avoid the damage'. This was removed by the Hague Protocol and only remains a possible defence in those jurisdictions that have adopted the Warsaw Convention but not the Hague Protocol.

Article 25A of the amended Warsaw Convention, clearly reverses this position by removing the 'negligent pilotage' defences and substituting provisions that basically bring the 'servant or agent' within the umbrella of the Convention limits. If action is brought against a servant or agent of the carrier arising out of damage to which the Convention relates they can avail themselves of the limits of liability available to the carrier under Article 22. These provisions are, however, strictly subject to the servant or agent acting within the scope of their employment and second, that damage did not result from their act or omission done with intent to cause damage or recklessly and with knowledge that damage would probably result, in which case the provisions of Article 25A shall not apply. This reflects the provisions of the original Article 25 discussed above.

Notice of damage

The time limitations on notification of damage can sometimes benefit the carrier where the unsuspecting passenger or consignor of cargo is caught out through ignorance of the provisions. Article 26 provides for notification of complaint in respect of baggage or cargo within certain time limits. Otherwise receipt by the person entitled to deliver the baggage or cargo (without complaint) is prima facie evidence that it is being delivered in good condition and in accordance with the document of carriage.

If damage occurs complaint must be made immediately. This should be made by 'the person entitled to delivery' to the carrier. At the very latest complaint must be made within the following time limits:

Baggage: within 7 days from the date of receipt
Cargo: within 14 days from the date of receipt

In the case of *delay* complaint must be made within 21 days from the date on which the baggage or cargo has been placed at the disposal of the person entitled to delivery. Provision is also made regarding the manner of complaint. It must be made in writing either on the document of carriage or as a separate notification despatched within the aforesaid time limit. Article 26 (4) states significantly that where a complaint fails to be made as provided within the time limits, no action shall lie against the carrier, save in the case of fraud on its part.

It is vital, therefore, that complaint be made within the time stated or (except in the unusual event of fraud on the part of the carrier) the complainant loses its right of remedy. In practice this can often work against consignors, and especially passengers who are ignorant of the Convention and its provisions. For example, seven days is not a long time for a typical holiday-maker who travels abroad for a minimum of 14 days on an annual holiday. Say, his luggage is damaged on the outbound trip and he decides to sort it out when he eventually gets back home. By the time he writes to complain 3 weeks might well have passed from the date he actually received the damaged luggage. Article 26 (4) would (in the absence of fraud) prevent him from claiming against the carrier.

In practical terms, these limits are very short, particularly in so far as they affect passengers and inexperienced consignors. After all it is not an area of law of which the average consumer has any real knowledge or awareness. It is surprising then that the present time limits have already been increased, as amended by the Hague Protocol. In the original Warsaw text the time limits were 3, 7 and 14 days in respect of baggage, cargo and delay, respectively. Indeed in recognition of the, perhaps unnecessary, severities of these provisions, reputable airlines do sometimes, as a matter of goodwill rather than legal necessity, consider claims made outside these time limits.

Chapter 3 then makes further provision as to the bringing of actions against a carrier.

Death of defendant

Where the person liable has died, Article 27 provides that an action for damages pursuant to the Convention lies against those who legally represent that person's estate. In practice, of course, the 'person liable', being the carrier, is unlikely to be an individual but rather a corporation.

Jurisdiction

The plaintiff bringing an action for damages under the Convention must bring the action in the territory of one of the High Contracting Parties which at the option of the plaintiff may be before the court having jurisdiction in one of the following territories:

(1) where the carrier is *ordinarily resident*, or
(2) where the carrier has its *principal place of business*, or
(3) where the carrier has an *establishment by which the contract has been made*, or
(4) at the *place of destination*

Questions of procedure are to be governed by the laws of the court hearing the case.

Time limits for action

Article 29 limits the time for bringing an action under the Convention to two years. If an action is not brought within this time the right to damages is extinguished. The two-year period is calculated as being two years from either:

(1) the date of arrival at the destination, or
(2) the date on which the aircraft ought to have arrived, or
(3) the date on which the carriage stopped.

The method of calculating this period is to be determined by the court hearing the case.

Successive carriage

Article 30 deals with successive carriage which occurs when the carriage is performed by successive or different carriers. When considering the definition of international carriage, Article 1 (3) specifically includes successive carriage within the Convention. It defines such carriage as that which is 'to be performed by several successive air carriers' which for the purposes of the Convention is to be deemed one undivided carriage so long as it is regarded by the parties as a single operation (see p. 168). Successive carriage should be distinguished from combined carriage which involves several different modes of transport and which is dealt with by Article 31 (see below).

Certain basic rules relating to successive carriers are set down in the three paragraphs of Article 30 and can be summarised as follows:

(1) Each (successive) carrier who accepts passengers, baggage or cargo is subject to the Convention and is deemed to be one of the contracting parties to the contract of carriage (in so far as the contract deals with that part of the carriage under its supervision).

(2) The passenger (or the passenger's representative) can take action only against the carrier who performs the carriage at the time during which the accident or delay occurred (except where, by *express* agreement, the first carrier has assumed responsibility for the whole journey).

(3) In respect of baggage or cargo, the passenger or consignor has the right of action against the *first* carrier, whereas the passenger or consignee who is entitled to delivery will have a right of action against the *last* carrier; each may take action against the carrier who performed the carriage during which the destruction, loss, damage, or delay took place. These carriers are jointly and severally liable to the passenger, consignor or consignee, as appropriate.

The Convention tries to cover the practical difficulties of claiming against the correct carrier: all too often damage or loss to goods will be apparent only at the end of the carriage and it will not be possible to discover exactly when it took place and which carrier is therefore responsible. In these circumstances the passenger or consignor can also take action against the first carrier (irrespective of fault) and likewise the passenger or consignee entitled to delivery, the last carrier. Without this protection a potential claimant, be it a passenger, consignor, or consignee, could have difficulty in pursuing a claim if it were unable to establish the appropriate defendant.

Combined carriage

Chapter 4 is the shortest of the five chapters of the Convention. It deals purely with combined carriage and consists of Article 31.

Article 31 defines combined carriage as that which is performed partly by air and partly by any other mode of transport. The combination aspect is, therefore, in respect of different modes of transport. This terminology should be distinguished from successive carriage, discussed above, which relates only to air carriage but performed by different carriers.

In the case of combined carriage, Article 31 (1), states that the Convention shall apply only to the carriage by air. This is, of course, wholly consistent with the aims and effect of the Convention which was never intended to regulate any other form of transport. Indeed, there are separate regulatory measures in many jurisdictions which cover other specific forms of transport. Article 31 (2), provides that nothing in the Convention shall prevent the parties in the case of combined carriage from inserting in the document of carriage conditions relating to other modes of carriage, provided of course that the provisions of the Convention are observed as regards the carriage by air.

Miscellaneous matters

Chapter 5 contains a number of general provisions and consists of Articles 32 to 40A.

Article 32

Article 32 confirms the essential force and effect of the Convention by declaring null and void any clauses in the contract or special agreements entered into before the damage occurred by which the parties will purport to infringe the rules of the Convention, whether by deciding the law to be applied or altering the rules as to jurisdiction. Despite the above, the Convention specifically allows clauses for arbitration in the case of carriage of cargo so long as these are subject to the Convention and the arbitration takes place within one of the jurisdictions referred to in Article 28.

Article 33

Article 33 states that nothing in the Convention shall prevent the carrier from either:

(1) refusing to enter into any contract of carriage, or
(2) from making regulations which do not conflict with the provisions of this Convention.

This makes it quite clear that, whilst the Convention is applicable to the extent that it so provides, the parties may enter into whatever other contractual terms they wish, so long as they do not affect the terms of the Convention. In practice, therefore, carriers are well advised to, and normally do produce additional and often extensive standard form conditions of contract which together with the statutory terms implied by the Convention, form the contract of carriage.

There are many useful areas that can be covered by such conditions of contract, for example, the relationship between the carrier and the passenger or consignor that exists before or after the period of time governed by the Convention. There are also many improvements that can protect the carrier in respect of the carriage of cargo, particularly problematic loads such as livestock. In that regard it should be remembered that there are particular regulations made in relation to the transit of animals. In the United Kingdom regulations are made and updated by statutory instrument such as the Transit of Animals (general) Order 1973 (SI 1973 No. 1377). These rules cover the carriage of animals by air and other modes of transport and provide for detailed provisions for the care and custody of animals. Any carrier likely to carry animals should be fully familiar with these provisions.

Article 34

This allows a rare exception to the otherwise essential provisions that relate to documents of carriage dealt with under Articles 3 to 9. These provisions do not apply, according to Article 34, in the case of carriage performed in extraordinary circumstances outside the normal scope of an air carrier's business. This would seem to provide for emergencies and other abnormal situations.

Article 35

This defines the expression 'days' as meaning current days, not working days, for the purposes of the Convention. In effect, therefore, when a period of days is referred to it must mean all days running consecutively within that period without the exclusion of any particular days.

Article 36

The original Convention was drawn up in French and, as already seen in Section 1 (2) of the Carriage by Air Act 1961, the French text shall prevail in the event of any inconsistency between that and the English. The English text is followed by the French text in Part II of Schedule 1 of the Act.

Article 36 confirms the Convention was drawn up in French and that a single copy of it shall remain deposited in the archives of the Ministry of Foreign Affairs in Poland. It also provides that a duly certified copy of it should be sent by the Polish government to the government of each of the High Contracting Parties.

Articles 37–40 deal with the coming into force of the Convention and such related technicalities (e.g. the process of ratification). It is worth noting, particularly in considering the USA position discussed in Chapter 34, that included under Article 39 is a process for denouncing the Convention. It states that any High Contracting Party may denounce the Convention by means of a notification addressed to the government of the Republic of Poland. The Polish government is then obliged to inform the government of each High Contracting Party. The date such denunciation is to take effect is to be six months after the notification is given. To conclude, therefore, any party to the Convention (being a High Contracting Party) may at any time give six months' notice to withdraw from the Convention. As we will see later (p. 214) the United States of America actually gave notice of denunciation in 1965 though this was later withdrawn.

Definition of 'state' and 'territory'

Article 40 deals with the inclusion or otherwise of a High Contracting Party's colonies, protectorates or territories. Article 40A was added by the Hague Protocol and in 40A (1) defines the term 'High Contracting Parties'. For the purposes of Article 37 (2), and Article 40 (1) the expression High Contracting Party is to mean 'state'. In all cases the expression is to mean a state whose ratification of, or adherence to, the Convention has become effective (and whose denunciation thereof has not become effective). In other words, the expression means states who are party to the Convention in that they have ratified it and are not merely original participants or subsequent signatories. This is a relevant distinction as, for example, some states signed the Warsaw Convention and/or the Hague Protocol but did not follow this with ratification. Most noticeably the USA signed the Hague Protocol but did not ratify it. It has not, therefore, officially adopted it within its jurisdiction where the original unamended Warsaw Convention remains in force.

Article 40A (2) clarifies the expression 'territory' which is to include not only the metropolitan territory of a state but also all other territories, the foreign relations for which that state is responsible. In the case of the United Kingdom these generally include territories such as Hong Kong, Bermuda and Gibraltar.

Future changes

Article 41 is the final designated Article of the Convention and provides the right for a High Contracting Party to call for the assembly of a new international conference to consider 'improvements'.

The English text ends with an 'Additional Protocol', a single additional provision relating to Article 2. Its purpose is to give the right to High Contracting Parties to exclude, if they so wish, international carriage performed directly by the state, its colonies, protectorates, or other territories.

Part II — the French text

Thereafter follows Part II of Schedule 1 of the Carriage by Air Act 1961 which incorporates the French text of the Convention, which as we have seen takes precedence in the event of a dispute as to meaning.

Conclusion

The Convention within the 1961 Act forms the basis of liability for the carrier in international carriage. It may not, of course, always be applicable. For example, the carriage in question must be within the definition of international carriage (thus the relevant states must be High Contracting Parties) and it must be for reward or otherwise by an air transport undertaking. Considering, however, that well over 100 countries have ratified the Warsaw Convention (and many of these the Hague Protocol) much of the world's international carriage by commercial undertakings is covered either by the Warsaw Convention or that Convention as amended by the Hague Protocol. It is, therefore, essential that an international operator is fully aware of the provisions and their effect. The actual text of the Convention should be studied carefully and professional advice sought as to further action that should be taken such as, for example, producing standard form conditions of carriage, which could be to the carrier's benefit.

Many High Contracting Parties, such as the UK, have adopted the Hague

Protocol, so it is more likely that the Convention applicable is that incorporating the Hague Protocol amendments. This is, of course, the text that we have been considering above as incorporated in the Carriage by Air Act 1961. It should always be borne in mind that whilst this is the version applicable in the United Kingdom, it may not be the version applicable in the jurisdictions of other High Contracting Parties. For instance, in the USA the version in force is that of the original Warsaw Convention of 1929 which contains a number of significant differences. It is, therefore, appropriate to look at the next stage in the Warsaw System, the Hague Protocol of 1955.

31 The Hague Protocol

It became apparent as early as the 1930s, within only a few years of the 1929 Warsaw Convention, that certain changes would have to be made. Most significant was the issue of quantity of damages, in terms of the value of the liability limits and how these amounts (expressed in Convention Francs) could be converted into the currencies of the individual High Contracting Parties.

Within ten years of the Warsaw Convention there were moves to make improvements. These were in the event held up by the Second World War but afterwards resumed, culminating in an international conference held in The Hague in 1955. This conference was convened by the International Civil Aviation Organisation (ICAO) and resulted in the Hague Protocol of 1955 which made a series of amendments to the existing text of the Warsaw Convention.

Where the Hague Protocol is applicable the actual text should be carefully studied. There are a number of changes, some affecting the liabilities of the parties and some of these amendments, by way of example, are listed below:

(1) *Terminology*
Certain key words are changed in the Hague Protocol, for example:

Warsaw (1929)	*Hague (1955)*
Airport	Aerodrome
Checked (baggage)	Registered (baggage)
Goods	Cargo
Transportation	Carriage

(2) *Simplification of documentation*
The Hague Protocol simplified some of the provisions and requirements relating to the documents of carriage.

(3) *Air waybill negotiability (Article 15)*
The Hague Protocol added a further provision, paragraph 3, to Article 15 which states, for the avoidance of doubt, that nothing in the Convention prevents an air waybill from being negotiable.

(4) *Negligent pilotage defences (Article 20)*
In the original Warsaw Convention there was a provision under Article 20 (2) which granted a potential defence to a carrier liable in respect of 'transportation' of goods and baggage if it could show that the damage was

occasioned by an error in piloting, the handling of the aircraft or in navigation (as discussed on p. 194 above). This defence was removed by the Hague Protocol with the deletion of paragraph 2.

(5) *Limit of liability (Article 22)*

The Hague Protocol deleted the Warsaw Article 22 and substituted a new Article 22, the main significance of which was to increase the limits of liability from 125,000 to 250,000 Convention Francs in respect of passengers. A new paragraph 4 was added to deal with legal costs and in effect to make limits of liability exclusive of legal costs (subject to certain qualifications).

(6) *Misconduct (Article 25)*

The Hague Protocol replaced the original Warsaw Article 25 with a new Article 25 that deleted the reference to wilful misconduct and subsituted different criteria: either an intent to cause damage; or recklessness with the knowledge that damage would probably result (see p. 192 above).

(7) *Servants'/agents' liability (Article 25A)*

The Hague Protocol also added a new Article 25A. This essentially extended similar protection, of the limits of liability, to the carrier's servants or agents.

(8) *Notification (Article 26)*

The Hague Protocol deleted the original Warsaw paragraph 2 of Article 26 and substituted a new provision, the main effect of which was to increase the notification period for complaint: from 3 to 7 days in the case of baggage, and from 7 to 14 days in the case of cargo. The notification period for delay was increased from 14 to 21 days.

Not all the states that ratified the Warsaw Convention also ratified the Hague Protocol, as already mentioned in the USA the unamended Warsaw Convention of 1929 still applies. The position of the United States of America is considered in more detail in Chapter 34.

32 The Guadalajara Convention

The next stage in the chronological development of the Warsaw System was the Guadalajara Convention which took place in the autumn of 1961 at Guadalajara, Mexico. The Guadalajara Convention dealt with a particular problem of definition within the Warsaw Convention: that of the term 'carrier'. Since 1929 practice had shown that it was not always clear who was the carrier or at least which carriers could potentially be included within that definition.

Article 1 (2) of the Warsaw Convention as amended by the Hague Protocol does not specifically define 'carrier', but, in expressing applicability in terms of carriage 'according to the agreement between the parties' interprets the carrier as essentially the contracting party. This potentially limits those carriers falling within the scope of the Convention to either the contractual carrier or the successive carrier as specified in Article 1 (3) and Article 30. Consequently an 'actual' carrier who was neither the contractual carrier or a successive carrier within those terms was outside the Convention. Most significantly such a carrier could not avail itself of the limitations of liability.

The purpose of the Guadalajara Convention is summed up in its title: 'Convention supplementary to the Warsaw Convention, for the unification of certain rules relating to international carriage by air performed by a person other than the contracting carrier'. The United Kingdom was an original signatory at Guadalajara and subsequently ratified the Convention which was incorporated onto the statute book by the Carriage by Air (Supplementary Provisions) Act 1962. This came into force on 1 May 1964 by virtue of the Carriage by Air (Supplementary Provisions) Act 1962 (Commencement) Order 1964 (SI 1964 No. 486).

The 1962 Act, though short, has a similar format to that of the 1961 Carriage by Air Act and contains the text of the Convention. It also contains some initial sections of a more general and administrative nature and then incorporates the text of the Convention in a Schedule, first in English and then in French. Section 1 (2) makes it clear that if there is any inconsistency between the two the French text is to prevail.

The English text consists of 18 Articles of which the last 8 deal with procedural matters mainly concerned with the coming into force of the Convention.

Article 1 specifies three essential definitions:

(a) the 'Warsaw Convention'
(b) the 'contracting carrier', and
(c) the 'actual carrier'.

The Convention prudently adapts the definition of 'Warsaw Convention' to include either the original Warsaw Convention of 1929 or as amended by the Hague Protocol in 1955, as relevant to the particular carriage at the time. Thus, in the UK the amended Warsaw Convention would now apply as incorporated in the Carriage by Air Act 1961.

Article 1 (b), defines the contracting carrier as the person who as principal makes an agreement for carriage governed by the Warsaw Convention (with a passenger, consignor, or person acting on their behalf). Article 1 (c), introduces the concept of the 'actual' carrier meaning

> a person, other than the contracting carrier, who, by virtue of authority from the contracting carrier, performs the whole or part of the carriage contemplated in paragraph (b) but who is not with respect to such part a successive carrier within the meaning of the Warsaw Convention. Such authority is presumed in the absence of proof to the contrary.

The liabilities of the actual carrier are then expressed in subsequent Articles.

Where an actual carrier performs the whole or part of the carriage governed by the Warsaw Convention (in accordance with Article 1 (b) above) both the contracting carrier and actual carrier (with certain exceptions) shall be subject to the rules of the Warsaw Convention; the former for the whole of the carriage contemplated in the agreement, and the latter for the carriage which it (actually) performs.

To conclude, in general terms, the 'actual' carrier is brought within the Warsaw Convention. It is liable for the carriage which it performs (whereas the contracting carrier remains liable for the whole carriage, originally contemplated).

Article 3 also deals with those acting as servants and agents, including in other words, employees of the actual carrier. The acts and omissions of the actual carrier and its servants and agents acting within the scope of their employment shall, in relation to the carriage performed by the actual carrier, be deemed to be also those of the contracting carrier. In addition, Article 3 (2) provides that the acts and omissions of the contracting carrier and of its servants and agents acting within the scope of their employment shall, in relation to the carriage performed by the actual carrier, be deemed to be also those of the actual carrier.

Various additional provisions clarify the 'actual' carrier's position as a carrier within the Warsaw Convention. For instance, Article 3 (2) provides that its liability is limited as specified under Article 22 of the Warsaw Convention.

Further provisions relate to notices and proceedings against the carrier. Complaints made to the carrier under the Warsaw Convention are deemed to be effective whether addressed to the contracting or the actual carrier. However, orders referred to in Article 12, thereof, shall only be effective if addressed to the contracting carrier (Article 4). An action for damages, in relation to carriage performed by the actual carrier may, at the option of the plaintiff (i.e. the person bringing the action) be brought against either the actual carrier or the contracting carrier or both (Article 7). Where, however, an action is brought against one such carrier it has the right to require the other to be joined in the proceedings (the procedure and effects being governed by the law of the court seized of the case) (Article 7).

For the avoidance of doubt, Article 10 expressly states that (except as provided in Article 7) nothing in the Convention shall affect the rights and obligations of the two carriers between themselves.

It should always be borne in mind that the purpose of the Guadalajara Convention is to bring the 'actual' carrier as therein defined, within the scope of the Warsaw Convention. It regulates, therefore, how this concept is fitted into the Warsaw framework. Some of these provisions have already been discussed. There are a number of miscellaneous provisions for consideration in the full text.

Conclusion

By now, it will be apparent to the reader that the Warsaw System whilst trying to maintain a logical and uniform development, has in its application developed a system of increasing variations, discrepancies and complexities. This was already apparent by 1961 when the Guadalajara Convention took place. The UK was, as we have already seen, a party to this and the previous two treaties (Warsaw and the Hague), all three of which are in force by virtue of being incorporated into UK legislation. Whilst the UK has been involved with each stage of the Warsaw development, other countries, however, have not. The Warsaw Convention was widely adopted but far fewer nations (a leading example being the USA) adopted the Hague Protocol and even fewer the Guadalajara Convention. Some adopted one or two but not the other or others. This selected adoption of other treaties has led to a hotchpotch pattern of applicable treaties with a variety of possibilities. This could be solely Warsaw, Warsaw amended by The Hague, either as amended by Guadalajara or not and so on.

After Guadalajara, by the mid-1960s, most states wanted further change but there was little agreement as to what this should be. The main bone of contention was the liability limits: whereas the concept of a liability limit was still acceptable to most, the amount of such liability was the cause of considerable disagreement and argument. Furthermore there remained the practical problem of how such a theoretical limit, once agreed, should be expressed and then converted into individual currencies. This conversion problem was becoming a fundamental impediment to the workings of the Warsaw System.

Indeed, it threatened its very existence as epitomised in the dilemma of the USA.

After Guadalajara (to which it was not a party) the USA in the 1960s looked to a satisfactory resolution of the monetary value of the liability limitation. It had been represented at The Hague but did not ratify or enforce that treaty essentially as a protest to the amount of the limits which it felt, particularly in the context of civil awards, were ridiculously and unacceptably low. It was unable to force the pace of international change at a speed that was acceptable to it and, therefore, took unilateral action in the form of the Montreal agreement of 1966 (which must be distinguished from the Montreal Protocol of 1975). The US position, will be reviewed in Chapter 34.

33 Post-Guadalajara

After Guadalajara there was increased dissatisfaction and debate about international aviation, resulting in further drafts and several international conferences which produced further treaties, as signed by a variety of different states. A lot of action took place but to little effect, at least as yet. Since Guadalajara in 1961, which was the last effective international treaty in the Warsaw System (and itself of limited effect in terms of substance and acceptance), there have been two major conventions resulting in five further protocols: the Guatemala City Protocol, 1971; and the Montreal Protocols, 1975, namely, additional protocols 1–4. So far none of these is in force (as at 1 January 1991). The UK has been a party and original signatory to each protocol and has gone as far as to pass legislation through parliament, notably the Carriage by Air and Road Act 1979, to give effect to certain of these provisions. However, though some parts of the Act are in force, mainly those affecting carriage by road, most provisions relating to the carriage by air have not yet come into force.

The Guatemala City Protocol 1971

We must now look at the next stage in the Warsaw System. Although not yet in force, it demonstrates the overwhelming desire for change even when agreement as to what those changes should be, is difficult to reach.

An international conference was held in Guatemala City, Guatemala, in 1971. The purpose was to agree a revision of the Warsaw Convention as amended at The Hague. The text put before the conference was basically produced by ICAO which was by now taking a leading role in trying to formulate an internationally acceptable 'new' Warsaw text. Over 50 states were represented though barely half signed the resulting protocol (including the UK and the USA). The Protocol is, however, not yet in force. The UK has made provision for ratification in the Carriage by Road and Air Act 1979, which incorporates certain amendments from Guatemala and subsequently Montreal. This Act, though passed, is not yet in force in so far as it relates to those provisions for the carriage by air.

Therefore only brief mention will be made of its contents.

Purpose

The heading of the Protocol states that it is a

Protocol to amend the Convention for the unification of certain rules relating to international carriage by air signed at Warsaw on 12th October 1929 as amended by the protocol done at The Hague on 28th September 1955.

Application

Article 1 merely states that the Guatemala Protocol shall apply modifications to the Warsaw Convention as amended at the Hague. Thus, the definitions of international carriage and applicability therein remain.

Liability

Most of the major changes found in the Guatemala text relate to the area of liability. For example, the original Warsaw Article 17 is deleted and replaced with various new provisions. The concept of liability is essentially that of strict liability placed on the carrier. It can still claim contributory negligence, though provisions under Article 21 are redrafted. The defence under Article 20 (1) of Warsaw, namely that the carrier may not be liable if it proves that it and its agents have taken all necessary measures to avoid damage has been deleted by Guatemala and Article 20 has been redrafted to restrict that possible defence.

Perhaps the most significant provision is the increase in the liability limits:

- Passengers   = 1,500,000 Francs
- Passenger delay = 62,500 Francs
- Baggage   = 15,000 Francs (per passenger)
- Cargo   = 250 Francs (per kilogram)

It will be noted that in the case of cargo the limit remains the same as that of Warsaw and The Hague.

Additional provisions, for example, include certain provisions allowing courts and jurisdictions where they would not otherwise be so authorised to award costs and the removal of provisions under Warsaw as amended by The Hague, allowing the maximum limits for passengers and baggage to be exceeded in the case of defective documentation.

Documentation

The Guatemala Protocol further simplifies passenger and baggage documentation. Notably, the requirement for the Warsaw/Hague notice is not necessary. The word 'ticket' is not used and a more flexible approach is adopted.

Significantly, non-compliance with certain documentary requirements within the substituted Article 3 (relating to passenger documentation) does not affect the liability limits as in the original provisions. There are similar changes for baggage redefined as 'checked baggage' instead of 'registered baggage'.

This gives an indication of some of the changes made by the Guatemala Protocol. For specific provisions reference should be made to the actual text. It should always be borne in mind that as at 1 January 1991 the Protocol is not generally in force. Indeed, it has in many ways been overtaken by subsequent events, namely the Montreal Protocols.

The Montreal Protocols 1975

The pressure for change and particularly some international resolution of the Warsaw dilemma was intense and a mere four years after the conference in Guatemala in 1971 a further international conference was held, this time in Montreal, Canada, in September 1975. Again it was convened under the auspices of ICAO which continued to play a key role in promoting and producing new solutions. The conference wanted in particular to deal with cargo and postal items and to tackle the continuing problem of liability limits and conversion into national currencies.

One of the main difficulties was the increasing discrepancies between nations as to which parts of the Warsaw System were in force in different jurisdictions. With this in mind, the Montreal conference produced a set of four protocols, the fourth dealing primarily with cargo and postal items, while the first three each amended a different version of Warsaw. None of these is yet in force (as at 1 January 1991). The UK has passed the Carriage by Air and Road Act 1979 which takes into account appropriate amendments made at Montreal, but this part of the Act is not yet in force. Thus, as with the Guatemala City Protocol, only a brief reference will be made to the substance of the Montreal Protocols.

Before referring separately to each protocol, it should be noted that one of the main objects of Montreal was to introduce a new monetary unit more easily convertible than the original Warsaw Francs which belonged to an age when most currencies were pegged to the gold standard. The new unit, in which limits of Additional Protocols 1, 2 and 3 are expressed, is the Special Drawing Right (SDR). This is a unit defined and valued by the International Monetary Fund and is easily converted into currencies of member nations. For non-member nations it is possible to determine their own method of conversion.

Additional Protocol No. 1

The first protocol amends the original version of the Warsaw Convention, 1929 (referred to in brackets for comparison purposes). It replaces the Warsaw limits expressed in Convention Francs with SDRs.

- Passengers   = 8,300 SDRs   (125,000 Convention Francs)
- Cargo/Registered =   17 SDRs per kg   (250 Convention Francs)
  baggage
- Objects of which =   332 SDRs   (5,000 Convention Francs)
  the passenger
  takes care

Additional Protocol No. 2

The second protocol follows closely the pattern of the first. It, however, amends the Warsaw Convention as amended at The Hague in 1955. Those limits (referred to in brackets for comparison purposes) are likewise replaced by SDRs.

- Passengers   = 16,600 SDRs   (250,000 Convention Francs)
- Cargo/Registered =   17 SDRs per kg   (250 Convention Francs)
  baggage
- Objects of which =   332 SDRs   (5,000 Convention Francs)
  the passenger
  takes care

Additional Protocol No. 3

The third protocol follows a similar pattern. It, however, amends the Warsaw Convention as amended at The Hague in 1955 and at Guatemala City in 1971. Those limits (similarly referred to in brackets for conversion purposes) are likewise replaced by SDRs.

- Passengers   = 100,000 SDRs   (1,500,000 Convention Francs)
- Delay in carriage =   4,150 SDRs   (62,500 Convention Francs)
  of passengers
- Baggage   =   1,000 SDRs   (15,000 Convention Francs)
- Cargo   =   17 SDRs per kg   (250 Convention Francs)

Additional Protocol No. 4

Whereas the Guatemala Convention amended certain provisions in respect of passengers and baggage (relating particularly to liability and documentation) the fourth Montreal Protocol aimed to tackle cargo and postal items. It amends the Warsaw Convention as amended at The Hague in 1955.

The Warsaw Convention, as amended by The Hague text (see p. 168), specifically states under Article 2 that 'this Convention does not apply to the carriage of mail and postal packages'. This is the version in force in the UK as incorporated in the Carriage by Air Act 1961 (see p. 162). The Montreal

Protocol No. 4 deletes the Hague Protocol Article 2 (2) and replaces it with a new qualified provision. It states that the provisions of the Convention (i.e. Protocol No. 4) shall not apply to the carriage of postal items, except as provided in the previous paragraph. That states that in the carriage of postal items (as they are now referred to instead of 'mail and postal packages') 'the Carrier shall be liable only to the relevant postal administration in accordance with the rules applicable to the relationship between the carriers and the postal administrations' (see Article II).

The other main area of reform dealt with by this Protocol is in relation to cargo both in terms of documentation and liability. Most of the provisions relating to the air waybill (Section 3 of Chapter 2 of Warsaw/Hague) have been deleted and replaced by new provisions. As with the changes made by the Guatemala City Protocol in respect of the passenger and baggage documentation, the air waybill requirements are to some extent simplified and the requirement for the Warsaw notice notably deleted. Significantly non-compliance with certain requirements of the air waybill no longer affects the rules of the Convention as amended in the Protocol, in particular the limitation of liability. This again extends the principle introduced by the Guatemala City Protocol in respect of passengers and baggage to include air waybills.

The Montreal Protocol No. 4 also follows the Guatemala City Protocol in applying what is essentially the principle of strict liability (though still subject to contributory negligence). Again, the old Article 20 (1) defence that the carrier, its servants and agents have taken all necessary measures is removed (in respect of cargo, except cases covering delay). It does, however, introduce some new defences under Article IV of the Protocol. It states that the carrier is not liable if its proves that the destruction, loss or damage to the cargo during the carriage by air resulted solely from one of the following four circumstances:

(a) inherent defect, quality or vice of the cargo;
(b) defective packing of that cargo performed by a person other than the Carrier or his servants or agents;
(c) an act of war or an armed conflict;
(d) an act of public authority carried out in connection with the entry, exit or transit of the cargo.

The provisions from paragraph 3 of the amended Article 18 are part of a complete redraft of Article 18 which replaces the Warsaw Article 18 which is deleted by Article IV of the Protocol.

As far as the monetary limits of liability are concerned, this is still based on the Warsaw Convention but is expressed in terms of Special Drawing Rights, thus unifying in SDRs all the monetary amounts under the Montreal Protocols. In this case, the limit is 17 SDRs per kilogramme (unless there has been a special declaration of value). This amount relates to the Warsaw amount of 250 Convention Francs.

This is an indication of some of the changes made by Protocol No. 4. As with the other protocols, reference to the text should be made for specific provisions. Again, it should be remembered that as at 1 January 1991 this Protocol is not in force.

Montreal in 1975 was the last of the major international conferences which produced amendments to the Warsaw System. The fact that the four protocols and the earlier Guatemala Protocol are not yet in force has placed in doubt the efficacy of further such conventions until there is a better chance of international agreement and resolution of the fundamental disputes over the limits and their conversion into national currencies.

34 The United States of America

Given its size and relative influence, the position of the USA has always been of great significance and its role within the Warsaw System affects the very efficacy of its continuance. In summary the historical position is thus.

The USA has, of course, adopted the original Warsaw Convention of 1929 which remains in force today in its unamended form. Although a prime force in promoting the reform that led to the Hague Protocol in 1955, that Protocol was never in fact ratified by the USA despite it being a signatory to the Protocol.

As already discussed in Chapter 32, it was not long after 1929 that the Warsaw Convention soon began to display some fundamental flaws in relation to the limit of the carrier's liability in terms of both amount and conversion into national currencies. The USA, in particular, with one of the most advanced systems of civil injuries claims was always a prime campaigner for higher limits. Even before the Second World War, its voice along with other conflicting views signalled the need for certain changes.

After the Second World War the debate continued and resulted in the international convention held at The Hague in 1955. Despite signing the protocol the following year, there was considerable disappointment that the changes had not gone far enough. Many felt the new passenger limits (increased from 125,000 to 250,000 Convention Francs) were far too low and some even felt that there should be no limits at all. Against these various voices of dissent, ratification and enforcement by the US government seemed increasingly unlikely.

This was not, however, a situation that could just be left in the air, so to speak. It also became imperative that the US position be resolved. It was increasingly obvious that the US could not force the international aviation community to accept higher limits at the level acceptable to the USA. It was insistent, however, that there were to be more acceptable measures in so far as it was concerned. In order to force the issue the USA took the drastic step in November 1965 of giving Notice of Denunciation of the Warsaw Convention. Such a procedure was provided for in Article 39 of the Convention whereby a High Contracting Party could at any time withdraw from the Convention by giving Notice of Denunciation to be effective at the termination of six months. In the event, the serving of the Notice had the desired effect and the Notice was subsequently withdrawn, thus leaving the USA as a High Contracting Party to the Warsaw Convention of 1929.

The giving of the Notice did achieve its purpose in concentrating international efforts to resolve the situation. In particular, ICAO convened a meeting in Montreal early in 1966. The debate focused on two conflicting viewpoints: that of the USA that the limits should be considerably increased, contrasted by the view that such increases would greatly prejudice many smaller airlines and states where civil injuries claims were far lower. In the latter case, not only would the claims be totally out of line with their own civil jurisdictions, but the cost of the additional insurance could be prohibitive, thus having a prejudicial effect against certain carriers and states. Unfortunately, no conclusive agreement was reached and the meeting in effect failed. The situation was, therefore, heading towards a crisis point.

It is true to say that the day was saved by the initiative of the airlines. Interested parties were by now anxious to ensure that the situation was speedily resolved. Many discussions took place between many different groups and interests but perhaps it was somewhat remarkable that resolution was a result of an agreement between airlines and not between governments or states (though it did receive governmental blessing).

The Montreal agreement

It was the initiative of the International Air Transport Association (IATA) (discussed in Chapter 1) which resulted in the Montreal agreement. Agreement was reached in May of 1966 literally only days before the US Notice of Denunciation ran out. This Notice was immediately withdrawn. The agreement was a compromise which established special limits of liability on carriage which affected the USA. This was to be effected by binding airlines (or in Warsaw terminology, carriers) who would be obliged to enter into the Montreal agreement. It is important to appreciate that the Montreal agreement is not an international treaty agreed between states and is not, therefore, technically part of the Warsaw System; it originated as an agreement between airlines and not states and as such is an example of private international law.

The original parties and signatories to the agreement were, therefore, a number of US and other international airlines whose services involved points in the USA. Subsequent carriers can become a party to (and therefore be bound by) the agreement, and there is now a list of several hundred signatories. A carrier may give notice to withdraw from the agreement. Though reached between airlines, the agreement has in effect the enforcement of the US government with whom it was originally filed and incorporated under the US Government Regulations (CAB No. 18900).

The text of the agreement contains certain basic regulations. It is applicable to carriers being the airlines who are signatories to the agreement. Where by reference to the contract of carriage, such carriage includes a point of origin, destination or agreed stopping place within the USA, a carrier is, within the context of the Warsaw Convention of 1929, or as amended by the Hague Protocol of 1955 (whichever is applicable in each case), to enter into a special contract in two significant respects. First, the liability for each passenger in respect of death, wounding or other bodily injury shall be the sum of either:

(1) US $75,000 inclusive of legal fees and costs or
(2) US $58,000 exclusive of legal fees and costs (for a claim made in a state where provision is made for separate award of legal fees and costs)

Second, the carrier shall not in respect of any claim arising out of death, wounding or other bodily injury of a passenger avail itself of the defence under Article 20 (1) of the Warsaw Convention of 1929 (or as amended in 1955).

Thus, not only are the limits of liability significantly increased but the carrier's main defence removed. Article 20 (1) states that the carrier shall not be liable if it proves that it and its agents have either (a) taken all necessary measures to avoid the damage or (b) that it was impossible for it or them to take such measures. This has the effect of leaving the carrier no Warsaw defence for passenger liability (in respect of physical harm) although it could still claim contributory negligence on the part of the plaintiff. The Montreal agreement itself also qualifies the removal of the Article 20 (1) defence by providing that the rights and liabilities of the carrier shall not be affected as against a person who has wilfully caused damage which resulted in death, wounding or other bodily injury of a passenger. The carrier has, therefore, been subjected to an increase in its potential liability and the removal of its main defence under the Warsaw Convention.

In line with the Warsaw Convention's documentary provisions the Montreal agreement also lays down certain regulations in respect of documentation. At the time of delivery of the passenger ticket each carrier is obliged to 'furnish' to each passenger the notice as specified. This notice in effect informs them of the special contract and significantly the Montreal agreement actually states the size of type and place of such notice. It is to be printed in type at least as large as 10 point modern type and in ink contrasting with the stock on:

1. each ticket;
2. a piece of paper either placed in the ticket envelope with the ticket or attached to the ticket; or
3. the ticket envelope.

The notice in CAB No. 18900 is generally as follows:

ADVICE TO INTERNATIONAL PASSENGERS ON LIMITATION OF LIABILITY

Passengers on a journey involving an ultimate destination or a stop in a country other than the country of origin are advised that the provisions of the treaty known as The Warsaw Convention may be applicable to the entire journey, including any portion entirely within the country of origin or destination. For such passengers on a journey to, from, or with an agreed stopping place in the United States of America, the Convention and special contracts of carriage embodied in applicable tariffs provide that the liability of [(name of Carrier) and certain other] carriers parties to such special contracts for death of or personal injury to passengers is limited in most cases to proven damages not to exceed US $75,000 per passenger and that this liability up to such limit shall not depend on negligence on the part of the carrier. For such passengers travelling by a carrier not a party to such special contracts or on a journey not to, from, or having an agreed stopping place in the United States of America, liability of the Carrier for personal injury to passengers is limited in most cases to approximately US $10,000 or US $20,000.

The names of carriers party to such special contracts are available at all ticket offices of such carriers and may be examined on request.

Additional protection can usually be obtained by purchasing insurance from a private company. Such insurance is not affected by any limitation of the carrier's liability under the Warsaw Convention or such special contracts of carriage. For further information please consult your airline or insurance company representative.

Note that subsequently a provision has been added to give notice of the US $58,000 limit exclusive of legal costs where appropriate. The above reference to insurance reminds us that one of the main reasons for this Montreal notice and the Warsaw and Hague notices is to advise passengers in advance of the potential limitations on the carrier's liability so that they have the opportunity to obtain their own insurance should they wish to do so. It should be noted that the above provisions relate to passengers only and do not apply to delay.

Administrative provisions provided that the Montreal agreement be filed with the then Civil Aeronautics Board (now with the Department of Transportation) for approval and other governments, as required. In effect, the agreement, originally an agreement reached between airlines then received governmental blessing and became enshrined in governmental regulations with the related powers of authority and enforcement.

Further regulations designated as CAB regulation ER837 of February 1974 require each carrier to include a further notice on each ticket in relation to baggage.

NOTICE OF BAGGAGE LIABILITY LIMITATIONS

Liability for loss, delay, or damage to baggage is limited as follows unless a higher value is declared in advance and additional charges are paid: (1) for most international travel (including domestic portions of international journeys) to approximately US $9.07 per pound (US $20.00 per kilo) for checked baggage, and US $400 per passenger for unchecked baggage; (2) for travel wholly between US points, to US $1250 per passenger on most carriers (a few have lower limits). Excess valuation may not be declared on certain types of valuable articles. Carriers assume no liability for fragile or perishable articles. Further information may be obtained from the Carrier, etc.

So far several hundred carriers, both US and foreign, have signed the Montreal agreement. Whether or not carriers intending to fly to or from points in the USA have already signed the agreement (and they normally have) it is almost certain that as a condition of their obtaining the necessary approval, normally in the form of a foreign air carrier's permit from the US authorities, they would be required to become a party to, and thereby accept, the Montreal agreement limits.

CAB Order (75–1–16)

In 1974 an order of the Civil Aeronautics Board made provision for an updated conversion of the Convention Francs into US dollars. This in effect provided a table of conversion rates (similar in function to the sterling equivalent orders in the UK (see p. 189) as well as citing certain other requirements in respect of liability limits. The conversion table specified is shown in Table 34.1, applicable as at 1st January 1991. It should be remembered that many carriers, particularly foreign international ones, will of course be subject to the rates applicable under the Montreal agreement and the terms of the above order.

Table 34.1

Conversion and Protocol Minimum Liability	Actual	Rounded
125,000 Francs (per passenger, Convention only)	$10,002.90	$10,000.00
250,000 Francs (per passenger, Hague Protocol only)	20,005.80	20,000.00
5,000 Francs (per passenger for unchecked baggage)	400.116	400.00
250 Francs (per kg for checked baggage and goods)	20.00580	20.00
250 Francs (per kg on a per-pound basis)	9.07460	9.07

Franklin Mint dilemma

No reference to the US position, however brief, could sum up its Warsaw dilemma without mention of one particular case: the Franklin Mint Case which was finally heard before the United States Supreme Court in April 1984. The issues involved the application of the Warsaw Convention (unamended version as applicable in the USA) limit of liability and, in particular, the conversion of that limit into the appropriate national currency, the US dollar. The problem was that no readily available method of calculation existed.

The facts of the case were essentially that there was a contract for carriage of cargo by the carrier, TWA, for Franklin Mint of four packages from Philadelphia to London. The packages were said to contain large quantities of valuable coins but no special declaration of value was made at the time of delivery. In the event none of the packages arrived in London (nor has since been found) and Franklin Mint consequently brought an action against TWA for the loss of the cargo. In that action Franklin Mint claimed a sum (approximately US $250,000) based on the alleged *actual* value of the contents of the packages. Procedurally the action went through three stages: a District Court; an Appeal Court; and, finally, the Supreme Court in Washington on the 17 April 1984.

The actual question of liability was fairly simple. It was accepted that TWA was liable under the Warsaw Convention. What was not clear was how this liability was to be evaluated in monetary terms. More specifically, how was the limit of 250 Convention Francs per kilogramme to be converted into US dollars? Four possibilities were discussed:

(1) Special Drawing Rights (SDRs) – the unit of currency used by members of the International Monetary Fund;
(2) the last official price of gold in the USA;
(3) the current exchange value of the French Franc; or
(4) the free market price of gold.

It was decided by the District Court at first instance that the second option, the last official price of gold, was the most applicable standard. On appeal, however, a decision of more far-reaching consequence was made, for the Appeal Court concluded that it had the authority to neither select one of the four options propounded, nor to create a new standard. It argued that as the judiciary, its authority consisted of interpreting existing laws and not creating new ones, which was the role of the legislature. In order to prevent immediate hardship and chaos it allowed the original judgment in this case to stand accepting the unit of calculation as the last official price of gold which should also apply generally for a period of 60 days. Thereafter that unit could not be used, nor any substituted; thus, at a stroke the Warsaw limits on liability for loss of cargo would in effect become unenforceable in the USA.

This led to much uncertainty and anxious speculation. Fortunately, the position was clarified and that judgment in effect reversed on appeal to the

Supreme Court which decided on 17 April 1984 that the Warsaw Convention cargo liability limits did remain enforceable and were not, therefore, rendered unenforceable in the United States. It confirmed that the rate of US $9.07 per pound approximating to US $20 per kilogramme was as good as any other. (It had been set by the Civil Aeronautics Board and was compatible with the Convention.) There was no obvious or binding evidence to suggest that the Convention no longer applied: no Notice of Denunciation had been given and it was certainly not up to the judiciary to repeal it. The Convention must be enforceable and remain so until changed by a government and meanwhile the court would uphold the above limits (calculated at the conversion rate of US $42.22 per ounce of gold).

Conclusion

The position of the USA remains that of compliance with the unamended Warsaw Convention which despite attempts to jeopardize its enforceability remains in force as held by the US courts. The USA has not, as we have seen, adopted the Hague Protocol, so, ironically, whilst arguing at the time that the liability limits were not high enough and consequently withdrawing its support in protest it has been left with the lower limits of the unamended Warsaw Convention. This means, for example, that the passenger limit is 125,000 Convention Francs (instead of 250,000 Francs under the Hague Protocol). However, this is, of course, academic in that the Montreal agreement of 1966, as we have seen, has provided for limits of up to US $58,000 (excluding legal costs) and US $75,000 (inclusive of legal costs) for applicable carriers and carriage. This was the United States' practical answer to the so far insoluble problems of the Warsaw Convention.

35 The Future

The Warsaw System still remains, with all its inherent problems and defects. None the less, its ideals are sound and its contribution to international aviation law profound. Its attempts to unify and clarify the regulation of international carriage have, in many ways, been successful for many years. It has established a framework of regulations adopted by most countries involved in international aviation and many of its provisions have given organisation and cohesion to vital areas such as the documentation of carriage and the fundamental principles of liability. There is no reason why on so many issues the Warsaw System should not continue so long as it can develop to incorporate changes when required by current demands.

We have seen that the only real threat to the 'system' is its one inherent flaw – the monetary limits of liability; the amounts and their conversion into individual currencies with the lack of general agreement on what these amounts should be. The use of SDRs would provide a solution to the problem of conversion but there still remains the problem of amount. The Montreal Protocols of 1975 attempted to solve the problems but it now seems unlikely they will receive general approval and acceptance in the near future, if ever. As the debate continues there is a growing opinion that believes there should, in fact, be no limit at all on the carrier's liability, at least in respect of passengers. This makes international agreement in the near future even less likely for most Third World countries and those with poor economies generally have argued against increases in the limits and would certainly not be able to finance unlimited claims. The cost of insurance premiums for such airlines would often prove prohibitive.

It should also be said that any such move towards unlimited liability is a fundamental departure from the essence of the Warsaw ideal upon which the whole system is based, that is, a balance of interests between the consumer and the airline. The one gained compensation on a no-fault basis whilst the other in return benefited from a ceiling or limit on its liability. This careful balance, though subject to some readjustment, has never actually been demolished. There have been, however, proposals to develop a new system based on absolute and unlimited liability on the part of the carrier.

To date various organisations, championing different interests, have made proposals for change ranging from promoting the implementation of the international conventions still not in force to the complete replacement of the

Warsaw System with a new system based on absolute and unlimited liability. Some proponents have produced drafts of amended or new systems (such as the International Law Association which based its proposals on absolute and unlimited liability). Meanwhile, ICAO, IATA and other aviation organisations continue to press for international co-operation and agreement on a unified system.

It can only be said in conclusion that the Warsaw System still survives and largely in its original form as created in 1929. That in itself is some achievement. It was perhaps ahead of its time then and after some 60 years of service still has much to commend it. Many of its provisions have provided a uniform practical framework for the carriage of passengers, baggage and particularly cargo. Despite discrepancies between jurisdictions over interpretation and enforcement and conversion of liability limits, many claims have been settled quickly, effectively and predictably without the need to resort to lengthy, expensive and unpredictable litigation. Apart from establishing a far easier and better system, particularly in the case of smaller claims, the system with its known limits has promoted easier resolution of disputes and resulted in cheaper insurance premiums payable by airlines which itself has kept down the cost of air travel to the consumer. The system has provided one significant and often underestimated commodity – *predictability* which in itself provides stability and a degree of fairness.

It should not be forgotten though that the actual Warsaw limits when applicable (whether those of the amended or unamended Convention) are, in practice, often superseded by higher limits as allowed by virtue of Article 22. Such a special contract can be entered into by a particular airline on a unilateral or voluntary basis. Many of the world's leading airlines have, in fact, adopted higher limits than the Convention in respect of passenger liability. An obvious example is the Montreal agreement of 1966 (see Chapter 34) whereby airlines enter into unilateral agreement to accept higher limits for flights involving the USA. Many individual airlines have, as a matter of policy, felt it appropriate to increase their limits.

In addition, the limits have been increased on a mandatory basis by governments and their aviation authorities. The CAA, for example, has certain standard form conditions it applies when granting route licences to airlines. One of these, presently known as Condition H, states:

> The licence holder shall enter into a special contract with every passenger to be carried under this licence on or after 1st April 1981, or with a person acting on behalf of such a passenger, for the increase of not less than the sterling equivalent of 100,000 Special Drawing Rights, exclusive of costs, of the limit of the Carrier's liability under Article 17 of the Warsaw Convention of 1929 and under Article 17 of that Convention as amended at The Hague in 1955.
>
> (*Official Record*, Air Transport Licensing, Series 1 p. 31
> (as updated November 1988)

Thus, one can again appreciate the web of confusion that faces the consumer of air transport. Quite apart from establishing whether or not the Warsaw Convention in its original, or amended, form applies one must establish what other provisions apply in relation to a particular country, carrier or route. This has confused the simplicity of the Warsaw ideal.

As to the future, it is difficult to predict. The Warsaw System has been subject to criticism for many years and yet seems no nearer abandonment or amendment now than at times in the past. This merely goes to make the point that there is no easy answer, particularly when trying to achieve international agreement. What has already been achieved by the Warsaw System should not be underestimated. Much of the system has worked well for many nations over many years. If only agreement as to new limits with an effective method of conversion, such as the use of Special Drawing Rights, could be reached, the Warsaw System should successfully survive for many years to come. Its only real threat is the proponents of unlimited liability, but this is unlikely to meet with general international approval in the foreseeable future given the enormous price that would have to be paid by airlines and/or governments in insurance premiums or other funding arrangements. The Warsaw System is, therefore, far from dead but it requires essential, but not extensive, surgery in order to give it a new life so that it can continue to give valuable service far into the future. It is indeed an asset of which the aviation industry should be proud; there are few such examples of constructive international organisation and agreement in the fields of law or diplomacy.

36 Non-international Carriage

So far we have looked in some detail at the position of international carriage, dominated as it is by the Warsaw System. There are, of course, a number of other categories of flight that are not covered by that definition, in particular, domestic flights but also other non-Convention carriage such as mail and postal packages (see Article 2 (2) of the Convention (Carriage by Air Act 1961)) and gratuitous international carriage not performed by an air transport undertaking (see Article 1 (1) of the Convention (Carriage by Air Act 1961)). In the UK legislation deals with these remaining categories of carriage under the Carriage by Air Acts (application of provisions) Order 1967.

The general purpose of the Order is, in effect, to wipe up non-Convention carriage. More specifically it applies to all carriage by air that is not carriage to which the amended Convention applies. The 'Amended Convention' is obviously the Warsaw Convention as amended at The Hague and is defined in the Order as the English text of such as set out in Schedule 1 of the Carriage by Air Act 1961, including the additional Article added after Article 41, known as the 'Additional Protocol' to the Warsaw Convention set out at the end of that Schedule. This includes a wide variety of carriage: here are some examples:

- Carriage of mail or postal packages
- Domestic carriage
- International carriage not within the definitions of applicability under the amended Convention (for example, where the places of departure/destination are not as appropriate, High Contracting States, or gratuitous carriage performed by a non-air transport undertaking)

The aim and general effect of the Order is to bring all carriage not already covered by the Carriage by Air Act 1961, i.e. non-amended Warsaw Convention carriage, within a similarly unified set of regulations.

Rather than create and apply a totally new or different code of rules, the Order sensibly accepts the basic ideas and provisions of the amended Convention text and applies them where appropriate with certain changes. The substance of the Order is, therefore, in effect, lists of details of specific amendments to the Convention text. For this purpose, the Order divides into two categories the carriage to which it applies. Schedule 1, applies to carriage

(see Section 4)

(a) which is not international carriage as defined in Schedule 2 of the Order, or
(b) carriage of mail or postal packages

International carriage as defined in Schedule 2 (see part A, Article 3 (2)) is basically the definition of the original, unamended Warsaw Convention of 1929. Thus Schedule 1 of the Order covers in effect all carriage which is neither within the amended or the unamended Warsaw Convention, i.e. non-Warsaw carriage. Schedule 2 of the Order covers carriage to which the Order applies, being carriage defined by that Schedule (see Section 5 (1)). As we have seen, that definition covers carriage to which the unamended Warsaw Convention of 1929 applies. Schedule 2 of the Order, therefore, deals solely with that category of carriage.

The balance of Section 5 and Sections 6, 7, 8 and 9 of the Order make appropriate references and amendments in respect of the 1961 Act and other legislation and Orders. Thereafter we are led straight into Schedule 1 followed by Schedule 2. Each Schedule is divided into various parts which detail its applicability to the appropriate Convention with specific amendments; the applicability of the Guadalajara Convention; and finally the amended draft of the appropriate Convention incorporating the changes made by this Schedule. Rather than repeat ad lib the list of omissions, amendments and other changes, the reader is best directed to specific conditions of the Schedules themselves.

By way of general comment and guidance it is worth noting that the aim of the Order is to try and create a system of unified and cohesive regulations applicable to carriage by air of passengers, baggage and cargo. The basic model is, in fact, the text of the Warsaw Convention with appropriate amendments according to the type of carriage. As will be seen from inspection of the Order, there are a number of changes, some apparently minor but all significant and thus each version of the regulations should be applied with care to the particular type of carriage.

The starting point must be first to establish whether or not a particular carriage is subject to the amended Warsaw Convention, and, therefore, subject to the Carriage by Air Act 1961. If not, the 1967 Order should be applicable, though it is then necessary to establish whether the carriage is subject to Schedule 1 or Schedule 2. It is easier to first establish whether Schedule 2 is applicable: is it carriage within the unamended Warsaw Convention? If not, Schedule 1 is almost certainly applicable, apart from rare exceptions such as carriage exempted by the Secretary of State. Article 8 of the Order empowers it to exempt any carriage or class of carriage or any person or class of persons from any of the requirements imposed by the Order, subject to such conditions as it thinks fit.

Part IV
The Environment

Protection of our environment is an increasingly important issue in today's world as the effects of pollution and past misuse become more apparent. The world of aviation is by no means immune from these concerns. Indeed, in the 1990s it is highly probable that environmental regulation will be the most significant issue affecting the aviation industry.

How then are airline operations affected by environmental considerations? To date by far the most significant impact on airlines has been as a result of aircraft noise. In terms of awareness this is hardly a new issue and, as we will see, attempts to restrict and regulate aircraft noise go back some thirty years to the 1960s. It has become a complex and vital issue with potentially serious and divisive consequences for the industry.

From today's perspective, with public awareness provoking increasing political action, we can foresee that other environmental issues will emerge. Already questions are being asked about atmospheric pollution caused by aircraft emissions, particularly on take-off and landing. Compared with noise this is a relatively new, unexplored and unregulated area. It will undoubtedly emerge, however, as the issue of the 1990s, as noise was the issue of the 1980s. In the second part of this section we will pursue the nature of this problem and its potential effect on the aviation industry.

The two areas of aircraft noise and emissions are the most obvious environmental problems for the industry but the environment is a wide, amorphous subject and in the final part of this section we will consider further environmental aspects relevant to aviation.

37 Noise

Introduction

Aircraft noise has been and remains a serious issue affecting the civil aviation industry. The key question is not whether there will be further noise regulations but when and how they will be introduced. The regulatory position at any one time can easily appear confusing with a plethora of different regulations in force and proposed at various levels. First, a clear distinction must be made between existing regulations and the various different proposals being suggested for the future. Second, the regulatory development can be clarified by source and type.

Regulations, actual or proposed, can originate at various levels: *internationally* by, for instance, political groupings such as the EEC or by aviation organisations such as ICAO; *nationally*, by legislation of an independent state; and at a *local* level by airports. Most airports have, of course, traditionally exercised some form of noise restriction with such measures as daily time curfews. Moreover, in the USA there have been in the past controversial moves (which pre-empted federal legislation) by local communities to inhibit on noise grounds certain airport operations.

Noise regulation falls into four basic types often introduced progressively, stage by stage. These can be summarised as follows:

(1) To stop noisy aircraft at their *inception* manufacturers can be controlled by restrictions on the certification of new designs (and new production versions of old designs).
(2) Once produced, individual nations can introduce restrictions on the *acquisition* of noisy aircraft by their airlines.
(3) One stage further is a total ban on the *operation* of those aircraft.
(4) Last, but not least, there is a variety of *local* rules at particular airports.

To summarise, noise rules can be seen to fall into a clearly defined pattern. They cover the *non-production*, the *non-addition*, and *non-operation* of noisy aircraft. They are complemented by *local* airports rules. In practice, they are also qualified by a series of exemptions.

Having set a basic framework for the source and type of regulations we can now look at what has actually been imposed and later consider the sort of proposals that are being made for the future.

Finally, it should always be remembered that the actual calculation and assessment of noise is not an easy task and apart from any difficulties in measuring sound the impact and effect on individual human beings is a subjective response. For example, take the case of loudly played pop music. To some this is sheer joy, to others it is deeply distressing. It has, therefore, been said that one man's noise is another's pleasure. One can assume that aircraft noise is not normally received as a pleasant experience. Even then, however, the tolerance levels of individuals will vary. To ensure some public protection it has been necessary, therefore, to establish some scientific basis to measure, assess and then apply definitive noise standards.

The main method of measuring noise is by means of perceived decibels, designated as PNdB. In very general terms this unit weighs different frequencies of noise. In practice this involves the establishment of standard measuring points on and near runways, where of course, the noise problem is best evaluated and controlled. Most major airports place microphones on long poles at measured distances which record the noise of aircraft taking-off and landing. From this information the normal noise levels emitted by particular types of aircraft can be measured at particular distances. These facts can be produced visually by showing noise contours, or 'footprints', as they are often known, of particular aircraft. In comparison terms an old 'noisy' aircraft, say a Boeing-707, would show a far larger footprint at certain levels of noise than, say, a modern, quiet-engined British Aerospace 146.

So noise can be tested and assessed. This is vital for the enforcement of mandatory standards which have been introduced over the years to restrict aircraft noise in vulnerable areas, in other words, on and near airports, where aircraft are taking-off, landing or in the process of so doing.

International background

The origin of existing noise regulations both in Europe and elsewhere is based upon the format established by ICAO in Annex 16 of its regulations. They include provisions for the uniform measurement of aircraft noise and the imposition of a standard to set maximum noise levels.

ICAO Annex 16 is in a sense the grandfather of noise regulations worldwide and it is an important starting point from which to trace national restrictions on different operations. The UK initiated the London Noise Conference ('International Conference on the Reduction of Noise and Disturbance Caused by Civil Aircraft') in 1966 when it became apparent that concerted international action was required on the noise issue. Although no positive action was taken then it led to further conferences culminating in the adoption of Annex 16 by the Council of ICAO on 2 April 1971. That original Annex (since designated Volume 1 − Aircraft Noise, of Annex 16 on Environmental Protection) remains in force, as amended from time to time.

Annex 16 essentially provides a format for compliance set out in three basic stages often referred to as Chapter 1, 2 and 3 as they relate to the actual Chapters 1, 2 and 3 respectively of the text of the Annex. In fact the Annex has a number of chapters and covers a variety of particular aircraft types or special circumstances (for example, Chapter 4 on supersonic aeroplanes and Chapter 8 on helicopters), but it is the first three chapters that lay down the vital general rules.

Chapter 1 of Annex 16 in fact deals with only administrative information though as a term 'Chapter 1' has been used to designate those aircraft that cannot comply with the standards laid down in Chapters 2 and 3. There are, therefore, three basic categories of noise standards.

First, the original unrestricted stage which will include aircraft which can not comply with Chapter 2 standards. These include most types of Boeing-707, McDonnell Douglas DC-8, Convairs, Caravelles and Tridents. While some have been re-engined or hush-kitted, most of these aircraft have now become obsolete in many parts of the world (for example, Europe and North America) because they do not comply with the noise regulations now in force.

Chapter 2 lays down the first level of noise standards and restrictions. In very general terms it applies to subsonic jet aircraft for which, before 6 October 1977, either:

(1) the application for certificate of airworthiness for the prototype was accepted, or
(2) another equivalent prescribed procedure was carried out by the certificating authority.

There are several exceptions relating to specific types of engine or, in particular, aircraft requiring a runway length, with no stopway or clearway of 610 metres or less at maximum certificated mass for airworthiness (this in effect exempts STOL aircraft, see p. 232). The aircraft to which Chapter 2 applies must comply with the standards specified. The noise evaluation measure is the effective perceived noise level, EPNdB, and the aircraft must, at certain specific points, not exceed the maximum specified EPNdB. Detailed technical data is provided in the Annex for the calculation of those standards.

Chapter 3 provides the second level of standards. It applies to all subsonic jet aircraft (including their derived versions) in respect of which an application for a certificate of airworthiness for the prototype was accepted (or another equivalent prescribed procedure carried out by the certificating authority) on or after 6 October 1977. It also exempts subsonic jet aircraft which require a runway length (with no stopway or clearway) of 610 metres or less at maximum certificated mass for airworthiness.

In addition Chapter 3 applies to certain categories of propeller-driven aeroplanes according to weight and date of airworthiness certification.

The noise standards of Chapter 3 are obviously stricter and in practice have the effect of excluding many more aircraft than those affected by Chapter 2 standards. For example, the next generation of civil jet aircraft is affected, such as most Boeing-727s, some Boeing-737s and early Boeing-747s.

The two levels of standards prescribed by Chapters 2 and 3 laid down a framework which has been the basis of nearly all noise regulations adopted or proposed world-wide. Many states, particularly in North America and Europe, have now introduced national regulations enforcing Chapter 2 type limits. Many are now working towards future Chapter 3 compliance.

Special cases

Annex 16 made special reference to different aircraft types and special cases. Most notable of these are *supersonic aircraft*, such as Concorde, and at the other extreme those aircraft having short take-off and landing capability, known as *STOL aircraft*.

Concorde

Concorde remains something of an anomaly though its use has been severely restricted by local opposition to many of its originally intended destinations. As far as Annex 16 is concerned, Chapter 4 dealing with supersonic aeroplanes in effect states that noise levels for the future production of existing SST-type aircraft must not exceed measured noise levels for the first certificated aircraft of that type. These are essentially those where the prototype was certificated before 1 January 1975 and clearly covers Concorde. At present no provisions are made within Annex 16, Chapter 4, for supersonic transport aircraft whose application for a certificate of airworthiness for the prototype was accepted on or after 1 January 1975.

Although Concorde will continue to be an anomaly, with no possibility whatsoever that it could comply with Annex 16, Chapter 2, it is unlikely that steps will be taken to prohibit altogether its operation for environmental reasons. This is not to say that it may not prove difficult in the future to obtain permission to operate Concorde to foreign countries on an ad hoc charter basis, a very profitable aspect of Concorde's operations in recent yers.

STOL aircraft

At the other extreme are those aircraft with the so-called STOL (short take-off and landing) capability. Annex 16 (in Chapters 2 and 3) specifically exempts aircraft requiring a runway length (with no stopway or clearway) of 610 metres or less at maximum mass for airworthiness. This, in effect, exempts STOL aircraft such as the Dash-7 used at London's Stolport, London City Airport.

In the UK such exemptions have been given statutory approval by virtue of the Civil Aviation Act 1982 which specifies the responsibilities of the CAA as further particularised in the Air Navigation (Noise Certification) Order 1990 S.I. 1990 No. 1514. Article 4 of that Order states that the provisions as to noise certification shall not apply to an aeroplane which in accordance with its certificate of airworthiness has a take-off distance required, at maximum total weight authorised on a hard level runway in still air in an International

Standard Atmosphere at sea level, of 610 metres or less. Thus, by echoing the original ICAO provision, STOL aircraft are exempted from noise certification legislation in the UK so long as they are capable of taking-off and landing on a runway of less than 610 metres.

This exemption has already been exploited by an innovative development in the UK. In the early 1980s plans were proposed to build a Stolport in London's declining dockland area. A planning application was submitted by the developers in 1982 and finally approved by the Secretary of State for the Environment in 1984. Meanwhile, there had been much debate and discussion, including vociferous local opposition. The main concern of local people was the amount of noise that would be created. Various local opposition groups were formed and support from further afield was sought.

Government approval may have quietened the noise protestors but not before a series of restrictions were imposed upon the Stolport operators. For instance, only Dash-7 or other Stol-type aircraft at least as quiet in all modes of operation (landing, take-off and on the ground) were to be allowed. When the CAA granted the original licences to two operators these were conditional on a number of specific restrictions. In brief, these affected the use, the type of aircraft, times of operations and total number of movements. For example, only Dash-7 type aircraft were to be used, a night curfew was imposed and no recreational flying was allowed.

The Stolport, now known as London City Airport, is in many ways the first purpose-built inter-city airport of its kind in Europe. It was a new venture exploring new concepts: frequent inter-city services with small aircraft taking the business passenger to and from the heart of the city environment with the maximum of ease in the minimum of time. Before this, airports by their very nature were constructed away from cities and built-up areas: the Stolport reverses this traditional philosophy.

A suitable site in terms of space, accessibility and general infrastructure requirements may be a rare commodity within an existing city environment. Once found any such opportunity is at the mercy of noise protestors. Noise is central to the operational philosophy of a Stolport: excessive restrictions could destroy it; it is a delicate balance.

The USA

One of the original objectives of governments, when the ICAO Annexes were discussed, was that there should not be a series of different rules covering noise certification in different parts of the world. There was, therefore, some disappointment when the USA decided to implement FAR 36 (Federal Airworthiness Regulations), rather than simply to accept the international standards of ICAO Annex 16.

Fortunately, in practice, the rules are similar (with a few differences) although their early implementation and application to foreign-registered aircraft had a disastrous effect on some airlines. The USA introduced an operational ban on Stage 1 aircraft (being in most cases aircraft that did not comply with ICAO Chapter 2 requirements) as from 1 January 1985. This included both home- and foreign-registered aircraft and few exemptions were granted. It should be noted that, whereas the ICAO standards are referred to as 'Chapters', the corresponding terminology of FAR 36 is 'Stages'. For more recent changes in the US see p. 241.

Europe

The position in Europe is now similar to that in the USA. Most European countries had banned aircraft not complying with Chapter 2 by 1 January 1988. However, many European countries allowed exemptions, at their discretion, until 31 December 1989.

For those countries in the European Economic Community, under the 1983 amended EEC Directive, member countries could grant exemptions to home-registered aircraft up until 31 December 1988 if those airlines agreed to replacement by Chapter 3 aircraft. These airlines, therefore, had to go from Chapter 1 to Chapter 3 in one step to qualify for that exemption. In relation to non-EEC registered aircraft, exemptions could be granted by each individual state beyond 1 January 1988 but only until 31 December 1989. The granting of these exemptions was, however, an entirely discretionary matter. Furthermore, in accordance with EEC Directives, exemptions would only be granted if the operator provided evidence of the economic and technical impossibility of continuing the services with Chapter 2 type aircraft.

Impact

The major flag carriers of Europe have already taken steps – sometimes painful – to configure their fleets to meet the present noise rules. Many of the aircraft recently disposed of, like Tridents and Caravelles, were not noise certificated and are not, therefore, available as a cheap resource for new airlines. However, many new start-up airlines in Europe are in the holiday charter sector and will tend to lease, not buy, aircraft. Older aircraft such as the frequently used Fokker-28s, Boeing 727-200s, McDonnell Douglas DC-9s, Boeing 737-200s, Airbus 300s, TriStars, McDonnell Douglas DC-10s and Boeing 747-100s, can meet Europe's Chapter 2 noise rules (if necessary with hush kits) – some meet Chapter 3 – and could be operated by start-up airlines. The primary scheduled airline opportunities in Europe following any liberalisation may well be for turbo-prop operators flying between regional cities, or from regional cities to hubs. Thus lack of cheap jets need not be an inhibiting factor in these circumstances. However, for airlines with an established fleet of non noise-compliant aircraft, the cost of capital investment in new aircraft could be very high.

Another point concerning the existing rules is that there were still a number of Boeing-707s and DC-8s on the European register in the mid-1980s. In some cases these have been hush-kitted or even re-engined. The problem is that near the end of their operational life these aircraft become increasingly inefficient and depreciate rapidly to a point where re-equipping becomes an essential and extremely expensive step. The worst problems were probably those suffered by cargo carriers, particularly the small, specialised cargo operators. In particular, these carriers suffered as a result of the following:

- competition in the cargo market from marginally priced belly-hold capacity provided by passenger widebodies;
- a realisation that the boom in cargo on the North Atlantic was likely to be short lived;
- the imposition of the US noise bans on foreign carriers (since 1 January 1985);
- the late availability of certificated hush kits;
- competition from non-European cargo carriers.

The all-cargo operator has, therefore, been particularly badly affected and it is no mere coincidence that the number of dedicated cargo-carriers in the UK has been greatly reduced during the 1980s. The majority of those operating in the late 1970s have not survived into the 1990s, primarily because they could not afford to re-equip their 'noisy' fleets with much newer, noise-compliant aircraft. Furthermore, many new cargo airlines, particularly in Africa, were emerging using Boeing-707s or DC-8s which were available cheaply on the second-hand market. These new airlines buy aircraft that are virtually obsolete in Europe and North America but, for a short time, they are able to take advantage of noise rules in Europe, until the expiry of their temporary exemptions, much to the disconsolation of home-based carriers. This, of course, came about because of the discrepancy in the imposition of Chapter 2 standards between home-based and foreign airlines in many European countries.

Despite, however, the fact that the carriers had this temporary advantage over home-based carriers, who had to comply earlier with Chapter 2 standards, there have been complaints of discrimination against Third World carriers whose non-compliant fleets are being banned from American and European destinations. Even allegations of economic imperialism have been made on the grounds that many less well-off nations can not afford to re-equip to meet the new noise standards and are, therefore, being pushed out of the market. These are merely some of the views expressed and complex issues raised by the imposition of noise restrictions.

The United Kingdom

Before reviewing recent noise regulations and those proposed for the near future, in the UK, it is worth considering the question of aircraft noise in the context of the general law.

Nuisance

To start at the beginning, the student of noise regulations should be aware that noise in general would normally be considered under the civil tort of 'nuisance'. Part III of this book reviews some basic areas of legal liability including civil wrongs or torts and, in particular, the tort of nuisance (see p. 150). It will be seen that whilst the common law provides a civil remedy under the tort of nuisance against a noise nuisance, statute law has intervened to eradicate such rights of action in respect of aircraft noise (and trespass) in most cases.

Section 76 of the Civil Aviation Act 1982 specifically states that

> No action shall lie in respect of trespass or in respect of nuisance, by reason only of the flight of an aircraft over any property at a height above the ground which, having regard to wind, weather and all the circumstances of the case is reasonable, or the ordinary incidents of such flight, so long as the provisions of any Air Navigation Order and of any orders under Section 62 above have been duly complied with and there has been no breach of Section 81 below.

In most cases, therefore, there is no right to bring an action in respect of a nuisance caused by aircraft noise. Such rights existed by virtue of the common law but have subsequently been removed by statute. The removal of this fundamental right, still available in many other jurisdictions, is the subject of increasing criticism here in the UK. Indeed, the position has been challenged in the European Court of Human Rights. However, in the case of *Powell and Rayner* (3/1989/163/219), it was held in a Judgment made in Strasbourg on 21 February 1991 that (inter alia) there was no violation of the Convention for the Protection of Human Rights and Fundamental Freedoms as claimed, in respect of Section 76. That Section remains effective and in force.

Noise protestors feel that without the backing of the law their effective strength has been considerably weakened. It is certainly true to say that the lack of legal weapons available to fellow protestors in some other jurisdictions, most notably in the USA, has made it particularly difficult in the UK. It is also true, however, that in the long term political means and public consensus can prove strong weapons. As far as noise restrictions are concerned, the UK is keeping up with the general level of governmental control adopted by other nations of similar standing.

Finally, before leaving the important provisions of Section 76 (1) it should be noted that the removal of the right is not absolute. There are exceptions (even if they be rare or difficult to prove in practice), for example, an aircraft flying so low that it is unreasonable in terms of the Act and/or it breaches the appropriate regulations. Other aspects of Section 76 are discussed in Part III (see p. 149–151).

We have, therefore, seen the creation by statute of an essential legal principle. This may seem heavily biased towards the aircraft operator and against the individual suffering from aircraft noise but it is not, however, a true representation of the policy of repeated governments. That policy has been to control, by increasing regulation, the amount of noise permitted by aircraft.

Measures have also been introduced to assist those living around airports, such as grants towards double-glazing for sound insulation.

Civil Aviation Act 1982

Looking again at the Civil Aviation Act 1982, there are many references throughout the legislation to the duties of the CAA in respect of aircraft noise, the making of noise regulations and so on. For example, after stating the general objectives of the CAA, in Section 4 (see p. 34), Section 5 states specific duties with regard to environmental matters. When exercising its aerodrome licensing function in respect of any applicable aerodrome, the CAA has a duty to have regard to the need to 'minimise so far as is reasonably practicable':

(1) any adverse effects on the environment, and
(2) any disturbance to the public

from noise, vibration, atmospheric pollution or any other cause attributable to the use of aircraft for the purpose of civil aviation.

Environmental issues can emerge in many different ways, for example, in relation to fixing aerodrome charges, aerodrome authorities are entitled to take into account aircraft noise (Section 38). For 'the purpose of encouraging the use of quieter aircraft and of diminishing inconvenience from aircraft noise' an authority may fix its charges by reference to inter alia:

(1) the amount of noise caused by the aircraft in respect of which the charges are made; or
(2) the extent or nature of any inconvenience resulting from such noise.

As we have seen in a number of regulatory matters, the power to give effect to the Chicago Convention and make regulations by means of an air navigation order is given by virtue of Section 60 of the Act. This states that an air navigation order may contain, amongst other things, provisions for:

> prohibiting aircraft from taking off or landing in the United Kingdom unless there are in force in respect of those aircraft such certificates of compliance with standards as to noise as may be specified in the Order and except upon compliance with the conditions of those certificates.

Further provisions, notably in Sections 77, 78 and 79, also provide for the making of noise-related regulations. For instance, under Section 77 an air navigation order may provide for regulating the conditions under which noise and vibration may be caused by aircraft on aerodromes. Importantly it states that no action shall lie in respect of nuisance by reason only of the noise and vibration caused by aircraft on specified aerodromes (in accordance with the provisions of an order made pursuant to that section). This potentially widens the general principles of Section 76 (1) in relation to overflying aircraft (see p. 236), to a general removal of a right of action in relation to aircraft noise and vibration at most aerodromes.

Under Section 78 the Secretary of State is given extensive powers to take various measures to restrict noise in relation to particular aerodromes, while Section 79 enables the Secretary of State to provide for grants towards the cost of insulation of buildings affected by aircraft noise.

Further to the extensive powers given by the Act to make regulations in respect of noise pursuant thereto, statutory instruments have provided such detailed regulations. The current regulations are found in the recent Air Navigation (Noise Certification) Order 1990 (SI 1990 No. 1514), as amended. This has essentially been updated from the previous Order of 1986 which incorporated the new regulations made by the EEC. As we have seen the EEC effected a ban (with a few temporary exemptions) on aircraft that did not in general terms meet Chapter 2 type requirements. Since then the temporary exemptions granted in specific cases have terminated and there is, in effect, a total ban within the UK and other EEC nations on the operation of aircraft which do not comply with the requirements of the EEC Directive. In other words, there is in most cases a general ban on the operation of non-noise certificated aircraft, being those that do not comply with the requirements of ICAO Annex 16, Chapter 2 type standards.

Recent developments

The next move is, therefore, to work towards Chapter 3 type compliance. It has long been accepted that the implementation of sudden, total bans on the operation of whole categories of aircraft would be quite unacceptable. Indeed, such measures would be rendered impracticable by their draconian effect on the civil aviation industry. To re-equip substantial fleets overnight would be impossible in economic and practical terms and the consequences for the industry and ultimately, therefore, the consumer would be disastrous. It has, therefore, long been accepted that noise regulation has to be progressive, giving reasonable notice in advance to airlines of future requirements so that they can plan in advance for the necessary changes which could, after all, involve them in substantial capital expenditure.

We saw in the introduction to this chapter that noise regulations could be categorised into various types and restrictions introduced in progressive stages before reaching an absolute total operational ban. The aim of the EEC Directive of December 1989 was to introduce measures to stop the *addition* of further aircraft that did not reach ICAO Annex 16, Chapter 3 requirements. In general, this Directive bans the addition to the registers of member states of civil subsonic jet aircraft which do not meet the noise requirements of Annex 16, Chapter 3 with effect from 1 November 1990. It does not, therefore, apply to aircraft which were already on the national registers of member states on that date. Nor does it apply to aircraft with a maximum take-off mass (i.e. weight) of 34,000kg or less and a capacity of 19 or less seats.

The overall practical effect of these measures is to exclude most aircraft of the following types: BAC 1-11; Boeings 707, 727, 737–100, 737–200; and

DC-8 and 9, unless any of these aircraft has been hush-kitted or re-engined to a Chapter 3 standard.

The Directive does provide a list of exemptions that can be obtained for certain categories of aircraft as exempted by member states. Examples of those exemptions within the Directive are as follows:

(1) aircraft of historic interest;
(2) aircraft used by an operator of a member state before 1st November 1989 under hire purchase or leasing arrangements still in effect and which in this context have been registered in a non-member state;
(3) aircraft leased to an operator of a non-member state which for that reason have been temporarily removed from the member state's register;
(4) an aircraft replacing one which has been accidentally destroyed and which the operator has been unable to replace by a comparable aircraft available on the market and certificated to Chapter 3 standard, provided that the replacement occurs within one year of the destruction of the original aircraft;
(5) aircraft powered by engines with a by-pass ratio of 2 or more.

In the UK the CAA has the power to consider and grant, if it considers appropriate, such exemptions. Under the EEC Directive any exemption as granted by member states must be notified to both the EEC Commission and to other member states. Clearly the purpose of the exemptions is to cushion any particularly harsh effects that the new regulations might impose on particular operators. One of the difficulties with introducing noise restrictions is to try and avoid measures that will be discriminatory and divisive in terms of a differing impact on individual operators. Obviously legislators try to avoid measures that would have a far worse effect on certain vulnerable operators.

There is also a second category of exemptions provided by the EEC Directive. These exemptions if granted by individual member states must be limited to a maximum initial period of three years, thereafter renewable for periods of up to two years, provided that all such exemptions must expire by 31 December 1995. There are two basic categories to which such exemptions can apply. They relate to the following:

(1) aircraft which are leased from a non-member state on a short-term basis provided that the operator demonstrates that this is the normal practice in his sector of the industry and that the pursuit of his operations would otherwise be adversely affected; or
(2) aircraft in respect of which an operator demonstrates that the pursuit of his operations would otherwise be adversely affected to an unreasonable extent.

These exemptions are clearly of a more subjective nature and the onus would be on an operator to demonstrate full compliance with these qualifications. The CAA has stated that only in exceptional circumstances will such exemptions be granted.

Following the 1989 EEC Directive, the UK brought into force the provisions of the Directive in the Air Navigation (Noise Certification) Order of July 1990, which came into force on 1 August 1990 (S.I. No. 1514 of 1990). The Order has already been amended and not only puts into effect the provisions of the EEC Directive but also deals in general with the requirements for noise certification. It applies to the following categories of aircraft:

(1) every propeller-driven aeroplane having a maximum total weight authorised of 9,000kg or less;
(2) every aeroplane which is capable of sustaining level flight at a speed in excess of Flight Mach 1.0, being an aeroplane in respect of which applicable standards are specified in Article 6 (9) of the Order;
(3) every microlight aeroplane;
(4) every other subsonic aeroplane which in accordance with its certificate of airworthiness has a take-off distance required, at maximum total weight authorised on a hard level runway in still air in an International Standard Atmosphere at sea level, of more than 610 metres;
(5) every helicopter, being a helicopter in respect of which applicable standards are specified in Article 6 (10) of the Order.

Subject to certain exceptions, aircraft to which the Order applies shall not land or take off in the UK unless there is in force a noise certificate issued by the CAA or other appropriate authority permitted by the Order.

The issuing authority is the CAA which must apply the standards and procedures specified in the Order. Once granted in respect of an aircraft registered in the UK, the noise certificate must be carried within that aircraft when in flight. Similarly an aircraft shall not land or take-off in the UK unless it carries any noise certificate which it is required to do so under the law of the country in which it is registered. The Order also provides for the production of noise certificates when requested by the CAA or other authorised persons. The CAA has extensive powers to suspend any noise certificate and if necessary, after due enquiry is made, to revoke, suspend or vary any such certificate, approval, exemption or other document.

More recently, in 1991, and most importantly, the EEC has produced proposals in the form of a draft Directive for an *operational* ban on aircraft that do not comply with Annex 16, Chapter 3-type standards. Whilst the Council of the EEC is due to act on these proposals before the end of 1991, it is unlikely that they will be agreed before the spring of 1992, to be effective before the end of that year.

In very general terms, the Directive proposes a progressive ban on the operation of certain aircraft that do not comply with Chapter 3-type standards from 1st April 1995. After that date those aircraft must cease to operate when they reach 25 years old or at the latest by 31st March 2002. This would apply to civil subsonic jet aircraft whose engines have a by-pass ratio of less than 2. It would not apply to aircraft with a maximum take-off mass of 34 000kg or less and a capacity of 19 or less seats. As usual a number of exemptions and rules for specific cases are proposed.

Finally, mention should be made of the latest US regulations published in September 1991. Having already introduced non-addition regulations these latest rules are the final step in achieving 'Stage' 3 compliance in the US by the end of 1999. They provide for progressive compliance levels to be met (by most civil subsonic aircraft) by the end of 1994, 1996 and 1998 with final compliance by the end of 1991. As usual, there are provisions for interim exemptions and waivers.

The future

The US has taken the major final step in providing for completion of its noise restriction programme by the year 2000. Elsewhere nations are at very different stages down this road. The EEC is, however, not far behind the US, having already made proposals that should produce final regulations in 1992. These are likely to provide for Chapter 3 compliance by 31st March 2002.

Meanwhile, it should also be remembered that there will no doubt be a progression of new measures to assist in reducing noise and its impact at airports. This is the area in which noise protesters are most influential. At most major airports in the UK there are established and well-organised protest groups. These groups are undoubtedly increasing their influence over politicians and the decision-making processes. While such groups have not yet developed their strength and influence to anything like that of their colleagues in America, the 1990s will no doubt see an increase in their activities in the UK.

It should not be forgotten that even with a relatively quiet voice back in the 1980s, protesters managed to stop on noise grounds the Heathrow–Gatwick helicopter link. Although the helicopter involved contributed only about 20 per cent of the total noise caused by helicopters along the route, the service was vetoed by politicians. Whilst this may have seemed a small matter to many at the time, it serves to demonstrate the strength of well-organised vociferous public opinion and was a warning to those airlines who choose to ignore such environmental issues. It will be interesting to see what opposition is raised to the recent proposals for an additional runway at Heathrow. Despite the fact that it is considered extremely desirable by many within the industry, it is 'already clear that the environmentalists' objections to such a development will be enormous. They will moreover be concerned with many more issues than that of aircraft noise, though this will no doubt be highest on the list'. (From 'Airlines, Tourism and the Environment', an article in *Tourism Management*, June 1991, by Stephen Wheatcroft, director of Aviation and Tourism International).

A decade or so ago some more short-sighted airline operators hoped, and indeed attempted to persuade politicians that noise restrictions were unnecessary and would prove extremely damaging to the industry. That is now, however, past history as many regulations are now well in force and most, if not all, airlines accept that they need to operate quieter aircraft. The only remaining question is when and how Chapter 3 standards will comprehensively be enforced.

38 Air Pollution

Introduction

Environmental pollution is a vast area of multifarious causes and effects that threatens the natural environment of our world. In broad terms, the natural world has three key elements: water, land and air. Within and between each of these is an interaction of functions and substances creating a natural order or balance, which sustains life on earth.

As we know only too well, the results of man's activities have in many ways interfered with and upset this balance. Today we see increasing evidence of the harm caused to the well-being of human, animal and plant life, which – if not controlled – will in the long term endanger the existence of our planet.

Air

Air is essential for life, and good quality air is necessary for the healthy existence of living organisms. Bad air affects all forms of life: people, plants and animals. Clothes suffer from dirty air, and it also affects property; the surface of buildings can be attacked and eroded by toxic pollutants.

Air can be affected by many substances, some visible such as dust, fog and smoke (not forgetting the notorious London smogs of the 1950s). Other more insidious substances are those invisible chemical gases with potentially dangerous characteristics. Air pollution can, therefore, be divided into two classifications: first, that caused by so-called particulate matter – for example smoke – essentially consisting of separate particles of matter and normally visible; and secondly, gaseous emissions which are, of course, invisible.

In many ways the former category relates to older, more traditional, industries, often involving the burning of fossil fuels (coal, oil, and gas). Some of these problems have already been tackled to some extent. For example, measures were introduced by the Clean Air Acts to stop the terrible London smogs that had reached horrendous proportions by the early 1950s; one significant improvement was to stop the burning of coal in domestic fires within London.

Gaseous pollutants have increased dramatically throughout this century, with the growth of an industrial economy and corresponding scientific developments. Unfortunately, the increasing sophistication of technology has been accompanied by an increase in the quantity and complexity of chemical pollutants. Before looking at specific pollutants relating to aircraft emissions, one or two general problem areas should be mentioned. Three of the most fundamental environmental problems are associated in some way with atmospheric pollution.

Greenhouse effect

First, the process of global warming referred to as the 'greenhouse effect'. This is essentially a detrimental condition caused by air pollution, which results in the continual warming of the earth's atmosphere. The eventual knock-on effect will be to melt ice-caps and raise the water level on the planet, thus causing flooding and the submergence of land beneath an expanding sea.

The warming is caused by an increase in certain chemicals, particularly carbon dioxide, which absorb and thereby trap the heat radiated from the sun. This can be compared to a greenhouse which allows the sun to shine through the glass and traps the heat inside. Through man's activities, the increased level of carbon dioxide – the most important 'greenhouse gas' – is allowing the earth's atmosphere to retain too much heat, thus the temperature is gradually rising. This gradual increase, even if only by a few degrees by the end of this century, could result in significant climatic changes. So whilst the existence of so-called 'greenhouse gases' such as carbon dioxide, nitrogen oxides, methane, and chlorofluorocarbons is essential for life, excessive amounts upset the natural balance and cause positively adverse effects.

Thus, whilst a certain amount of these gases is absorbed naturally (by the sea and plants) the excess is left in the atmosphere to absorb and retain heat that would otherwise have escaped. More specifically, carbon dioxide (in the right circumstances) is able to do this because it absorbs infra-red radiation. This is radiated from the earth, and that which is not absorbed (and therefore retained in the earth's atmosphere) goes back into space. The more carbon dioxide there is in the earth's atmosphere, the more heat is absorbed and therefore retained, increasing the temperature rather than escaping into space. In that sense the heat is being trapped by carbon dioxide.

Ozone layer

Secondly, there is the depletion of the ozone layer. Ozone is essentially three atoms of oxygen chemically bound to form a clear invisible gas. It is found in the earth's atmosphere approximately 10 to 30 miles above the ground and forms a vital protective layer around the earth. In particular, it shields the earth from ultraviolet light from the sun. As the sun's rays shine down on the earth, the ozone layer acts rather like a filter, stopping too many potentially-harmful rays from reaching the earth. Of course, some ultraviolet light does get through – otherwise, for example, people would not acquire a sun tan – but too much ultra-violet can be harmful to humans (for example, causing eye and skin problems) and to other forms of life.

The ozone problem arises because this important protective shield is being depleted, as a result of man producing certain chemicals to excess. Known as chlorofluorocarbons, this group of chemicals is used in the production and use of various everyday materials and appliances. Chlorofluorocarbons, for example, are found in many cooling appliances, from fridges to air-conditioning units, and are used extensively in aerosol sprays (though alternative propellants are now being introduced). They have proved extremely useful chemicals for the products of modern man but the hidden cost is now clear. Scientists have recently discovered that they are responsible not only for thinning the ozone layer, but creating actual holes within it. There is now considerable international concern at the potential danger to our planet if the ozone layer is further depleted.

Acid rain

Thirdly, there is the problem of acid rain, essentially rain containing polluting chemicals collected from the atmosphere. In particular, oxides dissolve in rainwater to form acids which, when deposited on the earth, can harm soil, plants, trees, fish in lakes and rivers, and buildings.

A common offender is the burning of fossil fuels: notably coal, gas and oil (including petrol in cars). These produce gases which are emitted into the atmosphere: mainly carbon dioxide and smaller quantities of nitrogen and sulphur oxides. These will eventually be absorbed by rain water forming corrosive acids – hence acid rain.

This rain can fall on neighbouring countries, causing harm to their lakes, rivers, and forests though they are not the polluting nation. The evidence can be dramatic: whole lakes devoid of fish, dying forests and decaying buildings. It is therefore particularly tragic when other nations deliberately destroy their rain forests, such as is happening in certain parts of South America, for example.

39 Aircraft-related Pollution

Aircraft emissions

Having established the nature and context of air-related pollution, we can now look more specifically at aircraft-related pollution and, in particular, aircraft emissions. An aircraft is essentially a fuel-burning machine, and like all such vehicles it emits or gives off certain chemicals that can pollute the atmosphere. First, we should consider the four main pollutants emitted from aircraft engines:

Carbon dioxide (CO_2)

Carbon dioxide, as we have seen, is the main contributor to the greenhouse effect. The man-made creation of excessive amounts of carbon dioxide is upsetting the previous natural balance so that all the carbon dioxide produced can not be absorbed naturally.

Carbon monoxide (CO)

Carbon monoxide is a product of inefficient combustion. It is toxic and particularly harmful to humans. As it is able to inhibit the blood's capacity to absorb oxygen, it can, in large enough quantities, cause death. (It is for this reason that the inhalation of car exhaust fumes, containing carbon monoxide, is sometimes used as a means of committing suicide.)

Nitrogen oxides (NOx)

Nitrogen oxides are another contributor to acid rain and the greenhouse effect. Amongst other things, they can also cause photochemical smog.

Hydrocarbons

Hydrocarbons are chemical compounds and another product of inefficient combustion. They include, for example, methane (a contributor to the greenhouse effect) and benzene (which can cause cancer in humans).

We can see, from the above examples, that aircraft engines emit chemicals, some of which (inter alia) are toxic as well as contributing to global warming, depletion of the ozone layer and acid rain: in so doing they are, in different ways, potentially harmful to our environment and ultimately plant, animal and human life.

The above examples may appear to paint a grim picture, but how serious in reality is the airline's contribution to atmospheric pollution? Clearly, airline operators cannot be blamed out of hand. As in so many examples of pollution from industrial and commercial activities, the harmful effects of such operations were unknown until recently, by which time those activities had become an accepted, indeed expected, normal part of modern life. It is hardly likely that once pollutants have been found or their dangers discovered, the activities out of which they arose would immediately cease – after all, no one would seriously suggest that all aircraft should be grounded.

The reaction is to undertake research and to adapt and improve engines and fuels in order to reduce the pollution. Of course, the older engines were the worst offenders, and today's engines are far more fuel efficient (which itself reduces the emission of such chemicals as hydrocarbons and carbon monoxide, produced largely as a result of inefficient combustion), quieter and generally cleaner. As technology improves, so will the engines.

It is also true to say that other improvements in the general operation of airlines could reduce emissions. For example, the recent problems with congestion and traffic control in Europe have resulted in many wasted hours of flying time, and aircraft queuing at low heights around airports waiting to land.

It has also been noted that an airport itself attracts traffic congestion on the ground. This, however, makes the point that perhaps puts aircraft emissions into context; it seems that aircraft emissions are relatively minor contributors to pollution compared with many other causes, notably the motor car, and it is widely accepted that the motor car is a serious pollutant of far greater significance than aircraft.

Having said that, all that can be done should be done to continually improve aircraft engines so as to reduce and eventually eliminate the emission of harmful pollutants.

40 Regulation of Aircraft Emissions

Awareness of the problems of aircraft emissions has been gradual and any regulatory response relatively slow in comparison with that of aircraft noise. There has been no singularly-identifiable event, such as the ICAO Annex 16 that addresses the problem and supplies solutions, or at least a means of solving the problem at a particular point in time. ICAO has, though, been instrumental in considering the emission problem.

A suitable place to begin is perhaps the United Nations Conference on the Human Environment held in Stockholm in 1972. ICAO was represented and its views recorded in the Assembly Resolution A 18-11. Amongst other things, the Resolution states

> in fulfilling this role ICAO is conscious of the adverse environmental impact that may be related to aircraft activity and its responsibility and that of its member states to achieve maximum compatibility between the safe and orderly development of civil aviation and the quality of the human environment.

In other words, this was a public statement by ICAO of both the awareness and responsibility of member states for such problems. In particular, member states must aim to strike a balance between the development of civil aviation and the quality of the human environment.

That Assembly adopted a Resolution (A 18-12) which called for the Council, with the assistance and co-operation of other international organisations, 'to continue with vigour the work related to the development of Standards, Recommended Practices, and Procedures and/or guidance material dealing with the quality of the human environment'. ICAO responded to these aims and soon established an Action Programme Regarding the Environment. As part of that programme a study group was established for the specific task of investigating aircraft engine emissions, the result of which was an ICAO circular entitled 'Control of Aircraft Engine Emissions' (Circular 134-AN/94) which was published in 1977. Various proposals were made and importantly a certification procedure was put forward for the control of vented fuel, smoke and certain gaseous emissions for new turbo-jet and turbo-fan engines intended for propulsion at subsonic speed.

ICAO's next move was to obtain more information, apart from technical data, in terms of consultation with the views of both experts and member states being sought. For this purpose a Council Committee, known as the

Committee on Aircraft Engine Emissions (CAEE) was established in 1977. It was at the second meeting of the CAEE in 1980 that important proposals were made for the incorporation of certain material into an ICAO Annex. Eventually this proposed material was adopted by the Council. It was decided that as another environmental issue it should form an addition to the original Annex 16. The original Annex 16 was, therefore, renamed 'Environmental Protection' and the original text which dealt solely with aircraft noise was reheaded 'Volume 1 – Aircraft Noise' and the new material on engine emissions became 'Volume 2 – Aircraft Engine Emissions'.

Volume 2 of Annex 16 now details the ICAO regulations in respect of engine emissions. As in the case of noise, it provides for the administration of certain regulations to apply to specified aircraft or engines with detailed calculations as to the assessment and measurement of the technical standards to be applied. Chapter 2 of Volume 2 provides for the applicability of these regulations to turbo-jet and turbo-fan engines as specified within that chapter. It relates to those only for propulsion at subsonic speeds and therefore excludes supersonic engines. It does not apply where certificating authorities make exemptions for specific engine types and derivative versions of such engines for which the type certificate of the first basic type was issued (or other equivalent prescribed procedure was carried out) before 1 January 1965. In such cases an exemption document must be issued by the certificating authorities.

The emissions involved, and therefore the types of gases to be controlled by certification, are as follows:

(1) smoke; and
(2) gaseous emissions, being – unburned hydrocarbons (HC)
                                   – carbon monoxide (CO)
                                   – oxides of nitrogen (NOx).

Detailed provisions are made as to the scientific calculations and procedures to be adopted in measuring such data and the standards to be applied.

Chapter 3 follows with regulations in respect of turbo-jet and turbo-fan engines intended for propulsion at supersonic speeds. It states that the provisions of the chapter shall apply to all such engines whose date of manufacture is on or after 18 February 1982. The emissions involved and to be controlled by certification of such aircraft are basically those named above as applicable to Chapter 2.

The requirements of Annex 16, Volume 2 in respect of aircraft emissions are by no means universally adopted by member states. Members are supposed to supply ICAO with updated information as to what regulations are in force in that state and any changes thereto. ICAO then circulates this information. To date, certain of these regulations are in force in the UK but the ICAO update, dated 1 October 1990, states that in respect of the UK controls have not yet been introduced to cover emissions of carbon monoxide and oxides of nitrogen. The reason given is that this is so that the UK keeps its regulations in line with those of the USA.

Within the UK the present regulations, so far as they exist in relation to aircraft engine emissions, are to be found in the Air Navigation (Aircraft and Aircraft Emissions) Order 1986 (SI 1986 No. 599) as amended. These regulations came into force on 1 May 1986. The Order applies to the following:

(1) every aircraft which is powered by gas turbine engines whose date of manufacture was on or after 1st May 1986 or in respect of which a certificate of airworthiness was first issued on or after 1 May 1986;
(2) every turbo-jet and turbo-fan engine whose date of manufacture was on or after 1st May 1986.

Essentially, the Order provides for the certification of the above aircraft and engines to comply with specified requirements. The requirements relate to:

(1) fuel venting, and
(2) smoke emission.

There are exceptions in that the requirements do not apply in certain circumstances, such as where an aircraft lands or takes-off on a prescribed flight and at a prescribed place for particular purposes (see Section 6).

To comply with fuel venting requirements, any aircraft to which the Order applies must not land or take-off in the UK unless there is in force an appropriate certificate of compliance in relation to the aircraft or the engines fitted to that aircraft. Such certificate must be issued by the appropriate authority, being the CAA in the United Kingdom, the competent authority of a contracting state (being any state which is a party to the Chicago Convention) or other competent authority of a country (applying standards approved by the Secretary of State).

To comply with the smoke emission requirements, an aircraft powered by turbo-jet or turbo-fan engines to which the Order applies must not land or take-off in the UK unless they are of a type which have been certified as complying with the smoke emissions requirements by a competent authority. A competent authority is the CAA in the UK or such authority abroad (as stated above with respect to the fuel venting requirements).

Significantly the requirements for obtaining a certificate relate essentially to smoke emission and do not as yet include emissions of carbon monoxide and oxides of nitrogen (as confirmed by the ICAO notice referred to above). Smoke is defined as 'the carbonaceous materials in exhaust emissions which obscure the transmission of light'. (See Section 2). These regulations are, therefore, seen as the beginning rather than the end of the story; it is likely that in the near future, further regulations will be made to include the control of additional chemical emissions and eventually to apply stricter standards in relation to specific toxic gases.

General legislation

There has in the past been various legislation which has touched in a variety of ways on environmental issues in the UK. Clearly the regulation of specific environmental problems is increasing as awareness and knowledge of the potential dangers become apparent.

In terms of atmospheric pollution, some of the earliest significant regulations were perhaps those made under the various Clean Air Acts introduced from the middle of the 1950s. These came about as a result of the severe smogs that were being suffered in urban areas, particularly London. Following that early legislation, such dramatic improvements were seen in the London air that the regular life-threatening smogs became a thing of the past.

Today the legislation continues, on a much wider basis. Of particular note is the Environmental Protection Act of 1990, which forms and updates much of the existing environmental legislation. It is neither the time nor the place here to investigate in any detail UK environmental legislation, but the latest Act should be seen as a watershed for the future. For example, it lays down some useful fundamental definitions. Section I of the Act defines the 'environment' as consisting of:

> All, or any, of the following media, namely, the air, water and land; and the medium of air includes the air within buildings and the air within other natural or man-made structures above or below ground.

Another essential definition is that of pollution being:

> The release (into any environmental medium) from any process of substances which are capable of causing harm to man or any other living organisms supported by the environment.

The Act provides for the making of further regulations and a framework of administration, inspection and control. It also gives detailed regulations on certain aspects such as waste and nuisances, as defined.

We will undoubtedly see further regulation of the environment in future years. Furthermore, the position here in the UK will also be influenced by our increasing involvement with the European Economic Community, since many environmental issues are seen as a potential area for European regulation and control.

41 Other Areas of Aircraft-related Pollution

In addition to aircraft emissions, it has to be said that in the multifarious operations of airline operators there are inevitably other areas of potential pollution. Some examples are given below:

1) Stripping, cleaning and de-greasing agents

Paint strippers, cleaning materials and de-greasing agents tend to contain pollutants. Some cleaning solvents and de-greasing agents, for example, contain methyl chloroform, which can cause ozone damage. There is evidence of some improvement, however, in the use of certain potentially harmful materials. For instance, some advances have been made in producing water-based paint strippers, which do not contain previously-used harmful chemicals. If these become widely used, there will be a significant improvement, given the enormous amount of paint that has to be stripped and removed on the world's aircraft.

2) De-icers

De-icing is obviously of vital importance every time aircraft take-off in certain weather conditions. However, de-icing agents may contain propylene glycol, a potential pollutant. Again improvements have been made, with a greater awareness of the problem leading users into taking action to ensure better and safer disposal of these chemicals.

3) Auxiliary power units

Such units can use kerosene, now regarded as a pollutant. Better units are presently being produced with alternative power sources.

4) Fire extinguishers

These can contain Halons which, in a similar way to chlorofluorocarbons, can damage the ozone layer.

5) General waste

Finally, one should perhaps mention the enormous potential for waste in aircraft operations. This can be wasted space, journeys and fuel (with unfilled aircraft, waiting periods and landing queues), wasted food and passenger consumables (from the food itself to plastic trays, cups, cutlery etc.) For example, the amount of plastic cups and containers and cutlery that is thrown away is enormous. This has caused some airlines to consider returning to metal cutlery that can be reused, or to the recycling of cups. Thus waste-saving programmes are beginning to be pursued, with some major airlines leading the way.

Quality of life

We have seen above some examples of the various areas of environmental vulnerability, particularly involving specific chemicals. There are other areas that are increasingly being viewed by people as generally damaging to the environment; we have already touched on the problems of waste (and, of course, noise in Chapter 37). It can also be said that airports, the centres of airline operations – apart from noise, chemical pollution and waste – cause congestion and tend to dominate the local environment. They can adversely affect property prices and intrude into the lives of all living creatures in the vicinity, from birds to human beings. Some regard them as a visual eyesore and there are indeed many planning and aesthetic issues connected with airports.

The future

Hopefully, we can all look forward to a better future whatever our role is, be it that of passenger, neighbour or airline. As science reveals the consequences of actions and substances, so it continues to find solutions, and improvements are being made all the time.

As far as aircraft operators are concerned, most are well aware that the environment cannot be ignored; good business sense dictates that airlines must be good neighbours or pay the consequences. Many operators recognise that in the long term it is commercial sense to be seen to clean up their act, and some of the major airlines have already taken steps to actively promote a cleaner and better environment. This may involve employing the necessary expertise to investigate problem areas, propose improvements and generally increase awareness within the airline of environmental considerations. A positive approach has also been adopted by IATA (the International Air Transport Association, see Chapter 1) by, for instance, the formation of an Environmental Task Force. There can also be direct commercial benefits involved in reducing certain pollutants, for example in the case of engine emissions, where '...the airlines are acutely aware of the need to reduce this problem and they have a strong financial incentive to do so because the harmful emissions are, for the most part, the products of inefficient combustion'. (From 'Airlines, Tourism and the Environment', *Tourism Management*, June 1991, by Stephen Wheatcroft, director of Aviation and Tourism International).

We have seen with the noise issue that, as public awareness increases, so does the active will to protest and cause change. Environmental issues in the 1980s emerged as a powerful new political issue, which is likely to play an increasing role in all our lives as we try to protect the world we have begun to abuse.

Appendices

Appendix I

The Freedoms of the Air

First Freedom

The right to OVERFLY i.e.; the right of an aircraft from State A to fly over State B without landing.

Second Freedom

The right to LAND i.e.; the right of an aircraft from State A to land and take-off in State B for technical purposes.

Third Freedom

The right to PUT DOWN i.e.; the right of an aircraft from State A to put down passengers, mail and cargo from State A in State B.

Fourth Freedom

The right to PICK UP i.e.; right of an aircraft from State A to pick up passengers, mail and cargo from State B (destined for State A).

Fifth Freedom

The right to carry traffic between points on a FOREIGN SECTOR i.e.; the right of an aircraft from State A to pick up passengers, mail or cargo in State B to put down in State C (and vice versa between State C and State B). In other words fifth freedom is a combination of 3rd and 4th freedom rights between points on a multi-sector route that are not in State A.

The above are the 5 principal freedoms of the air and assume that the aircraft and its operator are from State A. The 1st and 2nd freedoms are known as non-traffic freedoms and the 3rd, 4th and 5th freedoms as commercial or traffic freedoms.

Appendix II

Civil Aviation Authority

Civil Aviation Act 1982

PARTICULARS OF APPLICANTS FOR AND HOLDERS OF AIR TRANSPORT LICENCES

NOTES

1   Before completing this form, the applicant or licence holder should refer to the Civil Aviation Act 1982 (particularly Sections 64 – 70 and Sections 23 and 84), to the Civil Aviation Authority Regulations 1983 and to the Authority's Official Record, Series 1.

2   The form must be completed when first application for a licence is made, and on these occasions it will constitute part of the application as specified in Section 3.4 of the Official Record, Series 1. The attention of applicants is drawn to the fact that under Regulation 11(3) of the Civil Aviation Authority Regulations 1983 the Authority is required to make a copy of their application, including this form, available for inspection by the public.

3   Licence-holders should complete the form (on request from the Authority) at six-monthly intervals or when major changes have occurred. In these cases a copy will not be made available for inspection by the public unless the licence-holder gives its written consent to such disclosure. If no changes have occurred since the licence-holder's previous submission a written statement to that effect will be acceptable, provided that at least one full submission is made each year.

4   All questions should be answered or the words 'not applicable' entered. In some cases it may be necessary to attach a separate sheet for details where there is insufficient space. In the case of applicants or licence-holders which have a status other than that of a corporate body, the status should be specified after the name and other details given where applicable.

SECTION A   NAME, ADDRESS AND INCORPORATION

1   NAME (block letters)

---

2   Trading name if different from 1

---

3 Registered office	4 Address for correspondence
Telephone number	Telephone number

---

5   Date and place of incorporation

---

6 Company registration number	7 Date of financial year end

SECTION B   CAPITAL

1   Authorised share capital

---

2   Number and value of issued shares, indicating if any shares are not fully paid

---

3   If any shares have been issued other than for cash, state number and nature of consideration

---

4   If more than one class of share exists, give number of each class and details of each class

---

1

ATL 5
110584

SECTION C   SHAREHOLDERS

Where there are more than 20 shareholders in any company in sections C1, 2 or 3 below, details need be given only in respect of those holding more than 5% of the total shares issued. For this purpose nominee holdings should be counted with any shares held directly by the beneficial holder. The remaining shareholders should be grouped as 'other'.

1   Name in full and nationality of every shareholder giving number of each class of shares held and indicating in the case of nominee holdings the name and nationality of the beneficial holder

2   If a subsidiary of another company

   (i)   Name, address and place of incorporation of parent company

   (ii)   Name in full and nationality of every shareholder of parent company giving number and class of shares held, indicating in the case of nominee holdings the name and nationality of the beneficial holder

3   Name of ultimate holding company if different from that shown in C2 with other details as in C2 (i) and (ii)

SECTION D   SUBSIDIARY AND ASSOCIATED COMPANIES

1   Name and place of incorporation of any subsidiary companies, indicating proportion of shares held

2   Name and place of incorporation of any associated companies, indicating proportion of shares held or nature of association

SECTION E   CONTROL

Your attention is drawn to Sections 65(3) and 66(3) of the Civil Aviation Act 1982, which set out the Authority's duty if it is not satisfied that an applicant or licence-holder is either a United Kingdom national or a body which is incorporated under the law of the United Kingdom or of a relevant overseas territory or associated state and is controlled by United Kingdom nationals.

1   Give details of any person who or corporate body which has any significant financial interest in the applicant's or licence-holder's business (by way of shares, debentures, loans or otherwise) or can control the activities of the applicant or licence-holder in any way and is not such a person or body described above

SECTION F   DIRECTORS, MANAGEMENT AND STAFF

1   Name in full, position in company and nationality of each member of the board of directors

2   Name in full, position in company and nationality of other senior management

3   Experience of directors and senior management

In the case of new applicants, directors' and senior management personnel's air transport or other relevant experience should be stated. In the case of licence-holders this section need be completed only in respect of board members or senior management personnel appointed since the last form submitted.

4   Numbers of staff employed (giving maxima and minima if numbers fluctuate seasonally or otherwise)

    (i)   Aircrew including flight engineers   . . . . .

    (ii)   Cabin staff   . . . . .

    (iii) Ground engineering staff   . . . . .

    (iv) Other   . . . . .

         Total   . . . . .

I, the undersigned, hereby declare that to the best of my knowledge and belief the foregoing particulars are true and complete.

Signature

Signatory's name and official position

On behalf of   Date

Civil Aviation Authority

CIVIL AVIATION ACT 1982

APPLICATION FOR A SCHEDULED SERVICE (CLASS 1) LICENCE OR THE VARIATION REVOCATION OR SUSPENSION OF A CLASS 1 LICENCE

FOR OFFICIAL USE ONLY

Application No:

Date received:

Before completing this form the applicant should read carefully the following:

NOTES

1   Before making an application to the Authority, the applicant should refer to the Civil Aviation Authority Regulations 1983 and, in particular, to Section 67 (6) of the Act. The Authority is required to make a copy of each application available at their office for inspection by any person (Regulation 11 (3)) and the applicant may be required to serve a copy of the application on any person making an objection or representation (Regulation 15 (4)).

2   The original and four copies of this form should be sent to ERG Division, Civil Aviation Authority.

3   Attention is drawn to the requirement in Regulation 11 (2) that a copy of the application be served on the holder of the licence if he is not the applicant.

1   APPLICANT: (Name in full) ____

Address: _____

_____

Telephone No: _____

2   Is this application for a new licence or variation, revocation or suspension of an existing licence?

(i)   If the application is for a new licence complete Appendix 1 (Questions 9–13).

(ii)   If the application is for the variation of a licence complete Appendix 2 (Questions 14–19).

(iii)   If the application is for the revocation or suspension of a licence state:

(a)   Name of licence holder if not applicant:

(b)   Is the licence to be suspended or revoked?

(c)   Date from which it is proposed the revocation should be effective: _____

(d)   The period for which the suspension is required to be in effect,

from _____ to _____ inclusive.

3   Has the applicant satisfied himself that adequate aerodrome and other facilities, including customs and immigration are already available for use with the aircraft proposed at all places to be served?   YES/NO

If the answer is 'NO' state what the deficiencies are and when and on what authority they are expected to be remedied.

4   State whether the Air Operator's Certificate currently held permits the proposed service.   YES/NO.

5   Has the applicant satisfied himself about the effects of actual or potential airspace or airport constraints on the proposed service?   YES/NO.

6   Type of aircraft:

7   (i)   Brief indication of the existing or potential need or demand for the proposed service(s).

     (ii)   The proposed routeings of such services (ie, either as a single sector or a multi-sector route).

8   Particulars of any capital expenditure, financial commitment or commercial agreement, being particulars which the applicant as the holder of an air transport licence wishes the Authority to take into consideration in support of the application.

CERTIFICATE

I, the undersigned hereby apply for the grant of this application as described herein and I declare that, to the best of my knowledge and belief, the statements given in this application are true.

Dated this _____ day of _____ 19 _____

Signature _____ Position _____

Signatory's name in BLOCK LETTERS _____

On behalf of _____

APPLICATION FOR A SCHEDULED SERVICE (CLASS 1) LICENCE

9 Date from which licence is required to be in effect _____

10 State which of passengers mail or other cargo is to be carried _____

_____

11 Particulars of aircraft to be used on the service:

Aircraft type	Number		Capacity of each type
	Now available	*To be acquired*	

12 Sectors should be listed below in alphabetical order:

1 (Origin)	*2 (Destination & vv)*	*3 (Frequency or other restriction)*

Please specify the airport to be used (e.g. London/Heathrow or London/Gatwick)

13 Standard Provisions with Respect to Tariffs: I, VI, XI, XXIV.*

delete as appropriate
(continue on separate sheet if necessary)

APPLICATION FOR THE VARIATION OF A SCHEDULED (CLASS 1) LICENCE

14 Licence holder if not applicant:

15 Number of licence to be varied:

16 Date from which variation is required to be effective:

17 If details of existing sectors are to be varied, please state:

    (a) those terms and conditions of present licence which are to be varied:

    (b) revised terms and conditions sought:

    (c) the intended effect of the variation:

18 If new sectors are required they should be listed below in alphabetical order:

1 (Origin)	2 (Destination & vv)	3 (Frequency or other restriction)

    Please specify the airport to be used (e.g. London/Heathrow or London/Gatwick)

19 Standard Provisions with Respect to Tariffs: I, VI, XI, XXIV.*

*delete as appropriate
(continue on separate sheet if necessary)

Civil Aviation Authority

Air Navigation Order 1989

**APPLICATION FOR REGISTRATION OF AIRCRAFT
OR CHANGE OF OWNERSHIP**

FOR OFFICIAL USE

*This form, when completed, should be forwarded to the Civil Aviation Authority, Aircraft Registration, CAA House,
45–59 Kingsway, London WC2B 6TE, and must be accompanied by the appropriate registration fee (see Table B). The fee may be
paid by certain credit cards (see back of form), cheque or postal order (payable to the Civil Aviation Authority). A public counter is
open at this address between 1000 and 1600 hours, Monday to Friday.*

UNLESS THE FORM IS FULLY COMPLETED THE REGISTRATION OF THE AIRCRAFT WILL BE DELAYED.

1	(a)	Designation of aircraft: Name, Type, Series (as described by the Constructor)	(a)	
	(b)	Classification according to column 4, Table A	(b)	
	(c)	Number of engines fitted	(c)	
	(d)	Maximum total weight authorised (kg)	(d)	
	(e)	Charge payable (see Table B)	(e)	
	(f)	Specify if in microlight category (see Note vi)	(f)	
2	(a)	Name of Constructor and country of manufacture	(a)	
	(b)	Year of construction	(b)	
3		Aircraft Constructor's Serial Number		
4		UK Registration Mark Current	(a) G—	
		Proposed	(b) G—	
		Former	(c) G—	
5		Registration Mark and Nationality Mark if registered outside the United Kingdom and now being imported, or any Military Marks		

6	Person(s) or Body Corporate in whose name(s) the aircraft is to be registered (if necessary please continue on separate sheet) (See Note (i).)	Surname: Forenames: (in full)	Surname Forenames: (in full)
7	Permanent private address(es) of individual(s) or business address of Body Corporate in whose name(s) the aircraft is to be registered (please include post code). (See Notes (i) and (ii).) Telephone number at which applicant can be contacted during normal business hours.	 Post Code:	 Post Code:

CA 1   200291

8 If the aircraft is to be registered in the name of a Body Corporate, state:

(a) whether the aircraft is owned wholly by that body: (a) ....

(b) in what part of the Commonwealth that body is incorporated; (b) ....

(c) in what country is its principal business; (c) ....

(d) what is the address of its registered office; and (d) ....

(e) COMPANY REGISTRATION NUMBER; or date and place of registration where no number issued (e) ....

9 Is the person or body named in 6 overleaf (a) the owner of the aircraft or (b) a charterer by demise, i.e. by loan, hire or hire-purchase (but not a mortgage) agreement? *(Tick box as appropriate)*

If (b) above is applicable, state (i) the name and permanent address of the owner of the aircraft and (ii) the period of the charter.

☐ (a) Owner   ☐ (b) Charterer

(i)

(ii) ....

10 In what capacity is the person or Body Corporate in whose name the aircraft is to be registered entitled to have an aircraft registered in his name in the United Kingdom? *(Tick appropriate heading)*

☐ (a) The Crown in the right of Her Majesty's Government in the United Kingdom

☐ (b) British citizen.

☐ (c) Commonwealth citizen (see Note (iv)).

☐ (d) Citizen of the Republic of Ireland.

☐ (e) British protected person.

☐ (f) Body incorporated in the United Kingdom, or in some other part of the Commonwealth and having its principal place of business in the United Kingdom or in any other part of the Commonwealth.

☐ (g) Firm carrying on business in Scotland.

☐ (h) An unqualified person residing or having a place of business in the United Kingdom.

11 Is any unqualified person or body (other than 10 (a)—(h) above) entitled as owner to any legal or beneficial interest (other than as a member of a flying club) in the aircraft or any share therein?
If so, give particulars (name, address, nationality).

12 If the aircraft is to be registered in the name of an unqualified person residing or having a place of business in the United Kingdom (see 10 (h) above), state (a) the nationality of such person, and (b) the address of his residence or place of business in the United Kingdom, if not shown at 7 above.

(a) ....

(b) ....

I/We hereby declare that the foregoing particulars and answers are true in every respect, and I/We apply for the aircraft to be registered in the United Kingdom. (See Note (v).)

Date ....   Signature(s) ....
(Of all parties names overleaf — see Note (iii).)

Name(s) ....
(Block Letters)

Position held ....
(See Note (iii).)

TABLE A **GENERAL CLASSIFICATION OF AIRCRAFT ACCORDING TO PART A OF FIRST SCHEDULE TO AIR NAVIGATION ORDER 1989**
(See Section 1(b) of form)

1	2	3	4
Aircraft	Lighter than air aircraft	Non-power driven	Free Balloon / Captive Balloon
		Power driven	Airship
	Heavier than air aircraft	Non-power driven	Glider / Kite
		Power driven (flying machines)	Aeroplane (Landplane) / Aeroplane (Seaplane) / Aeroplane (Amphibian) / Aeroplane (Self-launching Motor Glider) / Rotorcraft (Helicopter) / Rotorcraft (Gyroplane)

TABLE B **CHARGES FOR THE ISSUE OF A CERTIFICATE OF REGISTRATION**
(See Section 1 of form)
Valid from 1 April 1991 until 31 March 1992:

	£
Balloon	29.00
Unpowered glider	29.00
Microlight aircraft	29.00
Any other aircraft having a maximum weight	
Not exceeding 750 kg	29.00
Exceeding 750 kg but not exceeding 15 000 kg	40.00
Exceeding 15 000 kg but not exceeding 50 000 kg	76.00
Exceeding 50 000 kg	95.00
* Out of sequence registration *(in addition to registration fee)*	
(a) in the case of a microlight aircraft	100.00
(b) in the case of any other aircraft	200.00
Charge for a replacement copy of a Certificate of Registration	10.00

See next page for details of payment by Visa/Access.

* Out of sequence registrations

Marks for microlight aircraft can be chosen from certain specified sequences, please contact Aircraft Registration for details.

Other aircraft may be allocated any four letter sequence which is either not in use, or has not previously been used. except for certain reserved groups of letters. Owners should note that registration marks cannot be transferred from one aircraft to another, although it is possible for an aircraft to be given a different set of marks subject to certain conditions.

NOTES TO BE READ WHEN COMPLETING THIS APPLICATION FORM

(i)   If the aircraft is to be registered in the name of an unincorporated body or more than one individual the full names and addresses of all persons sharing the ownership should be given. In the case of an unincorporated flying group, whose assets are held by trustees, the names and addresses of the trustees holding the assets should be given together with a separate list of the names and addresses of all members of the group. In the case of an aircraft chartered by demise and registered under the Air Navigation Order 1989, Article 4(5), the name and address of the charterer should be given.

(ii) A Club or hotel should not be given as the address unless the applicant resides there permanently.

(iii) **The owner or charterer should sign personally;** where more than one person is shown as owner (see Note (i)) each person should sign. In the case of a Body Corporate a Director, Secretary or other authorised officer of the company should sign, stating the position he holds, and a covering letter should be attached to the application giving the names of all such authorised officers.

(iv) The attention of applicants wishing to register aircraft is drawn to Section 37 of the British Nationality Act, 1981 which states that the phrase 'Commonwealth citizen' (see Section 10c) includes British dependent territories citizens, British overseas citizens and British subjects.

(v)   **The attention of all persons completing and signing this form is drawn to the importance of ensuring that the entries are correct. The making of a false statement for the purpose of procuring the issue of a Certificate of Registration is an offence under the Air Navigation Order 1989.**

(vi) 'Microlight' means an aeroplane having a maximum total weight authorised not exceeding 390 kg, a wing loading at the MTWA not exceeding 25 kg/square metre, a maximum fuel capacity not exceeding 50 litres and which has been designed to carry not more than two persons.

I wish to pay by Visa/Access, the charge specified in the current CAA Scheme of Charges, please charge to my account.

My card number is (13 or 16 Digits)

Expiry Date   /

Signature

Name on credit card (BLOCK CAPITALS)

Address (if not already given in Section 7)

Postcode

Civil Aviation Authority

The Mortgaging of Aircraft Order 1972

Register of Aircraft Mortgages

Mortgage Register Number

ENTRY OF AIRCRAFT MORTGAGE

PLEASE SEE NOTES ON REVERSE BEFORE COMPLETION

To be completed by Applicant:
I hereby apply for the mortgage, particulars of which are given below, to be entered in the Register of Aircraft Mortgages.

1 Date of mortgage.	
2 Description of the mortgaged aircraft (including its type, nationality and registration marks and aircraft serial number) and of any store of spare parts for that aircraft to which the mortgage extends. (The description of the store of spare parts must include an indication of their character and approximate number and the place or places where they are stored must be given. *(i)	
3 The sum secured by the mortgage. *(ii)	
4 Does the mortgage require the mortgagee to make further advances? If so, of what amount?	
5 Name and address and, where applicable, company registration number of the mortgagor.	
6 Register number of priority notice, if any.	

NOTES:

*(i) The description of the mortgaged property may, if necessary, be continued on a separate sheet, which shall be signed by the applicant.

*(ii) Where the sum secured is of a fluctuating amount, this should be stated and the upper and lower limits, if any, should be set out.

*(iii) Delete where inapplicable.

Signed ....

Name in block capitals ....

on behalf of *(iii) ....

....
(insert name and, where applicable, company registration number of mortgagee)

of ....

....

....
(insert address of mortgagee)

FOR OFFICIAL USE		
Persons Notified	Interest	Date

CA 1577
150980

NOTES FOR COMPLETION

1   This form should be completed and signed by the Mortgagee.

2   When completed it should be forwarded together with:

    (a)   the relevant fee and

    (b)   a certified true copy of the Mortgage deed;

    to the Civil Aviation Authority, Library, Records and Aircraft Register Services, CAA House, 45—59 Kingsway, London WC2B 6TE.

3   If the Mortgage covers more than one aircraft a separate entry should be made against each aircraft.

4   The Authority will confirm an entry in the Mortgage Register by returning a photocopy of the relevant entry to all parties named overleaf and to the current registered owner.

University of South Wales
Learning Resources Centre -
Treforest Self Return Receipt (TR1)

Title: Day & Griffin, the law of
international trade
ID: 731249089X

Total items: 1
19/06/2014 17:09

Thank you for using the Self-
Service system
Diolch yn fawr

Appendix III

Carriage by Air Act 1961 9 & IO Eliz. 2 c. 27

Note. — This Act, except for S. 1, post, and Sch. 2, post, came into force on 22nd June 1961, the date upon which it received the royal assent. Section I came into force on 1st June 1967: scc the Carriage by Air (Convention) Order 1967, S.I. 1967 No. 479, appendix C, post. For a full discussion on the effects of this Act see vol. I paras. 403–483. For text of Warsaw Convention, as unamended, see appendix A, ante.

This Act has been amended by thc Carriage by Air and Road Act 1979 to take account of the revision of the Warsaw convention as amended at The Hague 1955 made by certain protocols signed in Montreal on 25th September 1975. The 1979 Act is not yet fully in force and the amendments have been set out in footnotes where they are not yet operative.

Arrangement of sections

An Act to give effect to the Convention concerning international carriage by air known as 'the Warsaw Convention as amended at The Hague, 1955', to enable the rules contained in that Convention to be applied, with or without modification, in other cases and, in particular, to non-international carriage by air; and for connected purposes.

[22nd June 1961]

1. Convention to have force of law. — (1) Subject to this section the provisions of the Convention known as 'the Warsaw Convention as amended at The Hague, 1955' as[1] set out in the First Schedule to this Act shall, so far as they relate to the rights and liabilities of carriers, carriers' servants and agents, passengers, consignors, consignees and other persons, and subect to the provisions of this Act have the force of law in the United Kingdom in relation to any carriage by air to which the Convention applies, irrespective of the nationality of the aircraft performing that carriage, and the Carriage by Air Act 1932 (which gives effect to the Warsaw Convention in its original form), shall cease to have effect.

(2) If there is any inconsistency between the text in English in Part I of the First Schedule to this Act and the text in French in Part II of that Schedule, the text in French shall prevail.

(3) This section shall come into force on such day as Her Majesty may by Order in Council certify to be the day on which the Convention comes into force as regards the United Kingdom.

(4) This section shall not apply so as to affect rights or liabilities arising out of an occurrence before the coming into force of this section.

---

[1] The words 'further amended by provisions of protocols No. 3 and No. 4 signed at Montreal on 25th September 1975 and' are prospectively inserted by the Carriage by Air and Road Act 1979 Sch. 2 (not yet in force).

---

2. Designation of High Contracting Parties. — (1) Her Majesty may by Order in Council from time to time certify who are[1] the High Contracting Parties to the Convention, in respect of what territories they are respectively parties and to what extent they have availed themselves of the provisions of the Additional Protocol at the end of the Convention as set out in the First Schedule to this Act.

(2) Paragraph (2) of Article 40A in the First Schedule to this Act shall not be read as extending references in that Schedule to the territory of a High Contracting Party (except such as are references to the territory of any State, whether a High Contracting Party or not) to include any territory in respect of which that High Contracting Party is not a party.

(3) An Order in Council under this section shall, except so far as it has been superseded by a subsequent Order, be conclusive evidence of the matters so certified.

(4) An Order in Council under this section may contain such transitional and other consequential provisions as appear to Her Majesty to be expedient.

---

[1] The words, 'either generally or in respect of specified matters' are prospectively inserted by the Carriage by Air and Road Act 1979, Sch. 2 (not yet in force).

---

3. Fatal accidents. — References in section one of the Fatal Accidents Act 1846[1], as it applies in England and Wales, and [in Article 3 (1) of the Fatal Accidents (Northern Ireland) Order 1977][2], to a wrongful act, neglect or default shall include references to any occurrence which gives rise to a liability under Article 17[3] in the First Schedule to this Act.

---

[1] This includes a reference to the Fatal Accidents Act 1976: see Sch. 1, para. 2 to that Act, post.
[2] As amended by the Fatal Accidents (Northem Ireland) Order 1977, S.I. 1977 No. 1251, Sch. 1.
[3] The words 'Article 17 (1)' are prospectively substituted for the words 'Article 17' by the Carriage by Air and Road Act 1979, Sch. 2 (not yet in force).

---

4. Limitation of liability. — (1) It is hereby declared that the limitations on liability in Article 22[1] in the First Schedule to this Act apply whatever the nature of the proceedings by which liability may be enforced and that, in particular —

   (*a*) [...][2]
   (*b*) the limitation for each passenger in paragraph (1)[3] of the said Article 22 applies to the aggregate liability of the carrier in all proceedings which may be brought against him under the law of any part of the United Kingdom, together with any proceedings brought against him outside the United Kingdom.

(2) A court before which proceedings are brought to enforce a liability which is limited by the said Article 22 may at any stage of the proceedings make any such order as appears to the court to be just and equitable in view of the provisions of the said Article 22[4], and of any other proceedings which have been, or are likely to be, commenced in the United Kingdom or elsewhere to enforce the liability in whole or in part.

(3) Without prejudice to the last foregoing subsection, a court before which proceedings are brought to enforce a liability which is limited by the said Article 22[4] shall, where the liability is, or may be, partly enforceable in other proceedings in the United Kingdom or elsewhere, have jurisdiction to award an amount less than the court would have awarded if the limitation applied solely to the proceedings before the court, or to make any part of its award conditional on the result of any other proceedings.

(4)[5] The Minister of Aviation may from time to time by order made by statutory instrument specify the respective amounts which for the purposes of the said Article 22, and in particular of paragraph (5) of that Article, are to be taken as eqivalent to the sums expressed in franc which are mentioned in that Article.

(5) References in this section to the said Article 22[1] include, subject to any necessary modifications, references to that Article as applied by Article 25.

---

[1] The words 'and Article 22A' are prospectively inserted by the Carriage by Air and Road Act 1979 Sch. 2 (not yet in force).

[2] Section 4 (1) (*a*) was repealed by the Civil Liability (Contribution) Act 1978, Sch. 1, para. 5 (1).

[3] The words 'in paragraph (1)(a)' are prospectively substituted for the words 'in paragraph (1)' by the Carriage by Air and Road Act 1979 Sch. 2 (not yet in force).

[4] The words 'or Article 22A' are prospectively inserted by the Carrige by Air and Road Act Sch. 2 (not yet in force).

[5] Section 4 (4) is prospectively repealed by the Carriage by Air and Road Act 1979 Sch. 2 (not yet in force).

[4A. Notice of partial loss.[1] — (1) In Article 26 (2) the references to damage shall be construed as including loss of part of the baggage or cargo in question and the reference to the receipt of the baggage or cargo shall, in relation to loss of part of it, be construed as receipt of the remainder of it.

(2) It is hereby declared, without prejudice to the operation of any other section of this Act, that the reference to Article 26 (2) in the preceding subsection is to Article 26 (2) as set out in Part I and Part II of the First Schedule to this Act.]

---

[1] As inserted by the Carriage by Air and Road Act 1979 s. 2 (1). This section entered into force on 4th April 1979: see ibid., s. 2 (2).

---

5. Time for bringing proceedings. — (1) No action against a carrier's servant or agent which arises out of damage to which the Convention relates shall, if he was acting within the scope of his employment, be brought after more than two years, reckoned from the date of arrival at the destination or from the date on which the aircraft ought to have arrived, or from the date on which the carriage stopped.

(2) Article 29 in the First Schedule of this Act shall not be read as applying to any proccedings for contribution between [persons liable for any damage to which the Convention relates][1], ...[2].

(3) The foregoing provisions of this section and the provisions of the said Article 29 shall have effect as if references in those provisions to an action included references to an arbitration; and subsections (3) and (4) of [section thirty-four of the Limitation Act 1980][3] or, in Northern Ireland, subsections (2) and (3) of section seventy-two of the Statute of Limitations (Northern Ireland) 1958 (which determine the time at which an arbitration is deemed to be commenced), shall apply for the purposes of this subsection.

[1] The word 'tortfeasors' has been replaced; Civil Liability (Contribution) Act 1978, Sch. 1, para. 5 (2).
[2] The words omitted were repealed by the Limitation Act 1963, s. 4 (4).
[3] As substituted by the Limitation Act 1980.

6. Contributory negligence. — It is hereby declared that for the purposes of Article 21 in the First Schedule to this Act the Law Reform (Contributory Negligence) Act 1945 (including that Act as applied to Scotland), and section two of the Law Reform (Miscellaneous Provisions) Act (Northern Ireland) 1948, are provisions of the law of the United Kingdom under which a court may exonerate the carrier wholly or partly from his liability.

7. Power to exclude aircraft in use for military purposes. — (1) Her Majesty may from time to time by Order in Council direct that this section shall apply, or shall cease to apply, to the United Kingdom or any other State specified in the Order.

(2) The Convention as set out in the First Schedule to this Act shall not apply to the carriage of persons, cargo and baggage for the military authorities of a State to which this section applies in aircraft registered in that State if the whole capacity of the aircraft has been reserved by or on behalf of those authorities.

8. Actions against High Contracting Parties. — Every High Contracting Party to the Convention who has not availed himself of the provisions of the Additional Protocol at the end of the Convention as set out in the First Schedule to this Act shall, for the purposes of any action brought in a Court in the United Kingdom in accordance with the provisions of Article 28 in the said Schedule to enforce a claim in respect of carriage undertaken by him, be deemed to have submitted to the jurisdiction of that court, and accordingly rules of court may provide for the manner in which any such action is to be commenced and carried on; but nothing in this section shall authorise the issue of execution against the property of any High Contracting Party.

8A. Amendments consequential on revision of Convention

A new s. 8A is prospectively inserted by the Carrige by Air and Road Act 1979, s. 3 (1) (not yet in force) as follows:

'8A. Amendments consequential on revision of Convention. — (1) If at any time it appears to Her Majesty in Council that Her Majesty's Government in the United Kingdom have greed to a revision of the Convention, Her Majesty may by Order in Council [provide that this Act, the Carriage by Air (Supplementry Provisions) Act 1962 and section 5 (1) of the Carriage by Air and Road Act 1979 shall have effect subject to such exceptions adaptations and modifications] [1] as Her Majesty considers appropriate in consequence of the revision.

(2) In the preceding subsection 'revision' means an omission from, addition to or alteration

of the Convention and includes replacement of the Convention or part of it by another convention.

(3) An Order in Council under this section shall not be made unless a draft of the Order has been laid before Parliament and approved by resolution of each House of Parliament.'

---

[1] The words in square brackets are to be replaced by 'make such amendments of this Act, the Carriage by Air (Supplementary Provisions) Act 1962 and section 5 (1) of the Carriage by Air and Road Act 1979' (see Sch 2 para 1 of the International Transport Conventions Act 1983) when s. 9 of the 1983 Act comes into force.

---

9. Application to British possessions, etc. — (1) Her Majesty may by Order in Council direct that this Act shall extend, subject to such exceptions, adaptations and modifications as may be specified in the Order, to—

(*a*) The Isle of Man;

(*b*) any of the Channel Islands;

(*c*) any colony or protectorate, protected state or United Kingdom trust territory.

The references in this subsection to a protectorate, to a protected state and to a United Kingdom trust territory shall be construed as if they were references contained in the British Nationality Act 1948.

(2) An Order in Council under this section may contain such transitional and other consequential provisions as appear to Her Majesty to be expedient, and may be varied or revoked by a subsequent Order in Council.

10. Application to carriage by air not governed by Convention. — (1) Her Majesty may by Order in Council apply the First Schedule to this Act, together with any other provisions of this Act, to carriage by air, not being carriage by air to which the Convention applies, of such descriptions as may be specified in the Order, subject to such exceptions, adaptations and modifications, if any, as may be so specified.

(2) An Order in Council under this section may be made to apply to any of the countries or places mentioned in paragraphs (*a*), (*b*) and (*c*) of subsection (1) of the last foregoing section.

(3) An Order in Council under this section may contain such transitional and other consequential provisions as appear to Her Majesty to be expedient, and may confer any functions under the Order on a Minister of the Crown in the United Kingdom or on any Government or other authority in any of the countries or places mentioned in paragraphs (*a*), (*b*) and (*c*) of subsection (1) of the last foregoing section, including a power to grant exemptions from any requirements imposed by such an Order.

(4) An Order in Council under this section may be varied or revoked by a subsequent Order in Council.

(5) An Order in Council under this section shall not be made unless a draft of the Order has been laid before Parliament and approved by a resolution of

each House of Parliament:

Provided that this subsection shall not apply to an Order which applies only to the Isle of Man or all or any of the Channel Islands.

11. Application to Scotland. — In the application of this Act to Scotland —

    (*a*) there shall be substituted —

       (i) [...]¹;

       (ii) for any reference to a tortfeasor, a reference to a person who has been or might be held liable for loss or damage arising from any such act or omission;

       (iii) [...]¹;

       (iv) for any reference to the issuing of execution, a reference to the execution of diligence;

       (v) for any reference to an arbitrator, a reference to an arbiter; and

       (vi) for any reference to a plaintiff, a reference to a pursuer;

    (*b*) for section three there shall be substituted the following section —

'3. Fatal accidents. The reference in Article 17² in the First Schedule to this Act to the liability of a carrier for damages sustained in the event of the death of a passenger shall be construed as including liability to such persons as are entitled apart from this Act, to suc the carrier (whether for patrimonial damage or solatium or both) in respect of the death.';

    (*c*) in section five, subsection (1) shall have effect notwithstanding anything in section six of the Law Reform (Limitation of Actions, &c.) Act, 1954; and in subsection (3), for the words from 'and subsections (3) and (4)' to the end of the subsection there shall be substituted the words 'and for the purpose of this subsection an arbitration shall be deemed to be commenced when one party to the arbitration serves on the other party or parties a notice requiring him or them to appoint an arbiter or to agree to the appointment of an arbiter, where the arbitration agreement provides that the reference shall be to a person named or designated in the agreement, requiring him or them to submit the dispute to the person so named or designated.'

---

¹ As repealed by the Limitation Act 1963, s. 10 (5).
² The words 'Article 17 (1)' are prospectively substituted for the words 'Article 17' by the Carriage by Air and Road Act 1979, Sch. 2 (not yet in force).

---

12. Application to Northern Ireland. — In the application of this Act to Northern Ireland any reference to an enactment of the Parliament of Northern Ireland, or to an enactment which that Parliament has power to amend, shall be construed as a reference to that enactment as amended by any Act of that Parliament, whether passed before or after this Act, and to any enactment of that Parliament passed after this Act and re-enacting the said enactment with or without modification.

13. Application to Crown. — This Act shall bind the Crown.

14. Short title, interpretation and repeals. — (1) This Act may be cited as the Carriage by Air Act 1961.

(2) In this Act the expression 'court' includes (in an arbitration allowed by the Convention) an arbitrator.

(3) On the date on which section one of this Act comes into force the Acts specified in the Second Schedule to this Act shall be repealed to the extent specified in the third column of that Schedule:

Provided that, without prejudice to section thirty-eight of the Interpretation Act 1889 (which relates to the effect of repeals), this subsection shall not affect any rights or liabilities arising out of an occurrence before that date.

SCHEDULES
FIRST SCHEDULE THE WARSAW CONVENTION WITH THE AMENDMENTS MADE IN IT BY THE HAGUE PROTOCOL

Note.—The Carriage by Air and Road Act 1979, s. 1 (1), prospectively substitutes for this Schedule a new Sch. 1, which contains the English and French texts of thc Convention as amended by provisions of protocols No. 3 and No. 4 signed at Montreal on 5th September 1975. The new Schedule is not yet in force and is set out in the 1979 Act, post.

PART I THE ENGLISH TEXT
CONVENTION FOR THE UNIFICATION OF CERTAIN RULES RELATING TO INTERNATIONAL CARRIAGE BY AIR

CHAPTER I SCOPE — DEFINITIONS

ARTICLE 1

(1) This Convention applies to all international carriage of persons, baggage or cargo performed by aircraft for reward. It applies equally to gratuitous carriage by aircraft performed by an air transport undertaking.

(2) For the purposes of this Convention, the expression *international carriage* means any carriage in which, according to the agreement between the parties, the place of departure and the place of destination, whether or not there be a break in the carriage or a transhipment, are situated either within the territories of two High Contracting Parties or within the territory of a single High Contracting Party if there is an agreed stopping place within the territory of another State, even if that State is not a High Contracting Party. Carriage between two points within the territory of a single High Contracting Party without an agreed stopping place within the territory of another State is not international carriage for the purposes of this Convention.

(3) Carriage to be performed by several successive air carriers is deemed, for the purposes of this Convention, to be one undivided carriage if it has been regarded by the parties as a single operation, whether it had been agreed upon

under the form of a single contract or of a series of contracts, and it does not lose its international character merely because one contract or a series of contracts is to be performed entirely within the territory of the same State.

ARTICLE 2

(1) This Convention applies to carriage performed by the State or by legally constituted public bodies provided it falls within the conditions laid down in Article 1.

(2) This Convention shall not apply to carriage of mail and postal packages.

CHAPTER II   DOCUMENTS OF CARRIAGE
SECTION I — PASSENGER TICKET
ARTICLE 3

(1) In respect of the carriage of passengers a ticket shall be delivered containing:

(a) an indication of the places of departure and destination;

(b) if the places of departure and destination are within the territory of a single High Contracting Party, one or more agreed stopping places being within the territory of another State, an indication of at least one such stopping place;

(c) a notice to the effect that, if the passenger's journey involves an ultimate destination or stop in a country other than the country of departure, the Warsaw Convention may be applicable and that the Convention governs and in most cases limits the liability of carriers for deaths or personal injury and in respect of loss of or damage to baggage.

(2) The passenger ticket shall constitute *prima facie* evidence of the conclusion and conditions of the contract of carriage. The absence, irregularity or loss of the passenger ticket does not affect the existence or the validity of the contract of carriage which shall, none the less, be subject to the rules of this Convention. Nevertheless, if, with the consent of the carrier, the passenger embarks without a passenger ticket having been delivered, or if the ticket does not include the notice required by paragraph (1) (c) of this Article, the carrier shall not be entitled to avail himself of the provisions of Article 22.

SECTION 2 — BAGGAGE CHECK

ARTICLE 4

(1) In respect of the carriage of registered baggage, a baggage check shall be delivered, which, unless combined with or incorporated in a passenger ticket which complies with the provisions of Article 3, paragraph (1), shall contain:

(a) an indication of the places of departure and destination;

(b) if the places of departure and destination are within the territory of a single High Contracting Party, one or more agreed stopping places being within the territory of another State, an indication of at least one such stopping place;

(c) a notice to the effect that, if the carriage involves an ultimate destination or stop in a country other than the country of departure, the Warsaw Convention may be applicable and that the Convention governs and in most cases limits the liability of carriers in respect of loss or damage to baggage.

(2) The baggage check shall constitute *prima facie* evidence of the registration of the baggage and of the conditions of the contract of carriage. The absence, irregularity or loss of the baggage check does not affect the existence or the validity of the contract of carriage which shall, none the less, be subject to the rules of this Convention. Nevertheless, if the carrier takes charge of the baggage without a baggage check having been delivered or if the baggage check (unless combined with or incorporated in the passenger ticket which complies with the provisions of Article 3, paragraph (1) (c)) does not include the notice required by paragraph (1) (c) of this Article, he shall not be entitled to avail himself of the provisions of Article 22, paragraph (2).

SECTION 3 — AIR WAYBILL

ARTICLE 5
(1) Every carrier of cargo has the right to require the consignor to make out and hand over to him a document called an 'air waybill'; every consignor has the right to require the carrier to accept this document.

(2) The absence, irregularity or loss of this document does not affect the existence or the validity of the contract of carriage which shall, subject to the provisions of Article 9, be none the less governed by the rules of this Convention.

ARTICLE 6
(1) The air waybill shall be made out by the consignor in three original parts and be handed over with the cargo.

(2) The first part shall be marked 'for the carrier', and shall be signed by the consignor. The second part shall be marked 'for the consignee'; it shall be signed by the consignor and by the carrier and shall accompany the cargo. The third part shall be signed by the carrier and handed by him to the consignor after the cargo has been accepted.

(3) The carrier shall sign prior to the loading of the cargo on board the aircraft.

(4) The signature of the carrier may be stamped; that of the consignor may be printed or stamped.

(5) If, at the request of the consignor, the carrier makes out the air waybill, he shall be deemed, subject to proof to the contrary, to have done so on behalf of the consignor.

ARTICLE 7

The carrier of cargo has the right to require the consignor to make out separate waybills when there is more than one package.

ARTICLE 8

The air waybill shall contain:

(*a*) an indication of the places of departure and destination;

(*b*) if the places of departure and destination are within the territory of a single High Contracting Party, one or more agreed stopping places being within the territory of another State, an indication of at least one such stopping place;

(*c*) a notice to the consignor to the effect that, if the carriage involves an ultimate destination or stop in a country other than the country of departure, the Warsaw Convention may be applicable and that the Convention governs and in most cases limits the liability of carriers in respect of loss of or damage to cargo.

ARTICLE 9

If, with the consent of the carrier, cargo is loaded on board the aircraft without an air waybill having been made out, or if the air waybill does not include the notice required by Article 8, paragraph (*c*), the carrier shall not be entitled to avail himself of the provisions of Article 22, paragraph (2).

ARTICLE 10

(1) The consignor is responsible for the correctness of the particulars and statements relating to the cargo which he inserts in the air waybill.

(2) The consignor shall indemnify the carrier against all damage suffered by him, or by any other person to whom the carrier is liable, by reason of the irregularity, incorrectness or incompleteness of the particulars and statements furnished by the consignor.

ARTICLE 11

(1) The air waybill is *prima facie* evidence of the conclusion of the contract, of the receipt of the cargo and of the conditions of carriage.

(2) The statements in the air waybill relating to the weight, dimensions and packing of the cargo, as well as those relating to the number of packages, are *prima facie* evidence of the facts stated; those relating to the quantity, volume and condition of the cargo do not constitute evidence against the carrier except so far as they both have been, and are stated in the air waybill to have been, checked by him in the presence of the consignor, or relate to the apparent condition of the cargo.

ARTICLE 12

(1) Subject to his liability to carry out all his obligations under the contract of carriage, the consignor has the right to dispose of the cargo by withdrawing

it at the aerodrome of departure or destination, or by stopping it in the course of the journey on any landing, or by calling for it to be delivered at the place of destination or in the course of the journey to a person other than the consignee named in the air waybill, or by requiring it to be returned to the aerodrome of departure. He must not exercise this right of disposition in such a way as to prejudice the carrier or other consignors and he must repay any expenses occasioned by the exercise of this right.

(2) If it is impossible to carry out the orders of the consignor the carrier must so inform him forthwith.

(3) If the carrier obeys the orders of the consignor for the disposition of the cargo without requiring the production of the part of the air waybill delivered to the latter, he will be liable, without prejudice to his right of recovery from the consignor, for any damage which may be caused thereby to any person who is lawfully in possession of that part of the air waybill.

(4) The right conferred on the consignor ceases at the moment when that of the consignee begins in accordance with Article 13. Nevertheless, if the consignee declines to accept the waybill or the cargo, or if he cannot be communicated with, the consignor resumes his right of disposition.

ARTICLE 13

(1) Except in the circumstances set out in the preceding Article, the consignee is entitled, on arrival of the cargo at the place of destination, to require the carrier to hand over to him the air waybill and to deliver the cargo to him, on payment of the charges due and on complying with the conditions of carriage set out in the air waybill.

(2) Unless it is otherwise agreed, it is the duty of the carrier to give notice to the consignee as soon as the cargo arrives.

(3) If the carrier admits the loss of the cargo, or if the cargo has not arrived at the expiration of seven days after the date on which it ought to have arrived, the consignee is entitled to put into force against the carrier the rights which flow from the contract of carriage.

ARTICLE 14

The consignor and consignee can respectively enforce all the rights given them by Articles 12 and 13, each in his name, whether he is acting in his own interest or in the interest of another, provided that he carries out the obligations imposed by the contract.

ARTICLE 15

(1) Articles 12, 13 and 14 do not affect either the relations of the consignor or the consignee with each other or the mutual relations of third parties whose rights are derived either from the consignor or from the consignee.

(2) The provisions of Articles 1, 13 and 14 can only be varied by express provision in the air waybill. (3) Nothing in this Convention prevents the issue of a negotiable air waybill.

ARTICLE 16

(1) The consignor must furnish such information and attach to the air way-bill such documents as are neccssary to meet the formalities of customs, octroi or police before the cargo can be delivered to the consignee. The consignor is liable to the carrier for any damage occasioned by the absence, insufficiency or irregularity of any such information or documents, unless the damage is due to the fault of the carrier or his servants or agents.

(2) The carrier is under no obligation to enquire into the correctness or sufficiency of such information or documents.

CHAPTER III LIABILITY OF THE CARRIER

ARTICLE 17

The carrier is liable for damage sustained in the event of the death or wounding of a passenger or any other bodily injury suffered by a passenger, if the accident which caused the damage so sustained took place on board the aircraft or in the course of any of thc operations of embarking or disembarking.

ARTICLE 18

(1) The carrier is liable for damage sustained in the event of the destruction or loss of, or of damage to, any registered baggage or any cargo, if the occurrence which caused the damage so sustained took place during the carriage by air.

(2) The carriage by air within the meaning of the preceding paragraph comprises the period during which the baggage or cargo is in charge of the carrier, whether in an aerodrome or on board an aircraft, or, in the case of a landing outside an aerodromc, in any place whatsoever.

(3) The period of the carriage by air does not extend to any carriage by land, by sea or by river performed outside an aerodrome. If, however, such a carriage takes place in the performancc of a contract for carriage by air, for the purpose of loading, delivery or transhipment, any damage is presumed, subject to proof of the contrary, to have been the result of an event which took place during the carriage by air.

ARTICLE 19

The carrier is liable for damage occasioned by delay in the carriage by air of passengers, baggage or cargo.

ARTICLE 20

The carrier is not liable if he proves that he and his servants or agents have taken all necessary measures to avoid the damage or that it was impossible for him or them to take such measures.

ARTICLE 21

If the carrier proves that the damage was caused by or contributed to by the negligence of the injured person the court may, in accordance with the provisions of its own law, exonerate the carrier wholly or partly from his liability.

ARTICLE 22

(1) In the carriage of persons the liability of the carrier for each passenger is limited to the sum of two hundred and fifty thousand francs[1]. Where, in accordance with the law of the court seised of the case, damages may be awarded in the form of periodical payments the equivalent capital value of the said payments shall not exceed two hundred and fifty thousand francs[2]. Nevertheless, by special contract, the carrier and the passenger may agree to a higher limit of liability.

(2)—(a) In the carriage of registered baggage and of cargo, the liability of the carrier is limited to a sum of two hundred and fifty francs[3] per kilogramme, unless the passenger or consignor has made, at the same time when the package was handed over to the carrier, a special declaration of interest in delivery at destination and has paid a supplementary sum if the case so requires. In that case the carrier will be liable to pay a sum not exceeding the declared sum, unless he proves that that sum is greater than the passenger's or consignor's actual interest in delivery at destination.

(b) In the case of loss, damage or delay of part of registered baggage or cargo, or of any object contained therein, the weight to be taken into consideration in determining the amount to which the carrier's liability is limited shall be only the total weight of the package or packages concerned. Nevertheless, when the loss, damage or delay of a part of the registered baggage or cargo, or of an object contained therein, affects the value of other packages covered by the same baggage check or the same air waybill, the total weight of such package or packages shall also be taken into consideration in determining the limit of liability.

(3) As regards objects of which the passenger takes charge himself the liability of the carrier is limited to five thousand francs[4] per passenger.

(4) The limits prescribed in this Article shall not prevent the court from awarding, in accordance with its own law, in addition, the whole or part of the court costs and of the other expenses of the litigation incurred by the plaintiff. The foregoing provision shall not apply if the amount of the damages awarded, excluding court costs and other expenses of the litigation, does not exceed the sum which the carrier has offered in writing to the plaintiff within a period of six months from the date of the occurrence causing the damage or before the commencement of the action, if that is later.

(5) [5]The sums mentioned in francs in this Article shall be deemed to refer to a currency unit consisting of sixty-five and a half milligrammes of gold of millesimal fineness nine hundred. These sums may be converted into national currencies in round figures. Conversion of the sums into national currencies

other than gold shall, in case of judicial proceedings, be made according to the gold value of such currencies at the date of the judgment.

---

[1] The words '16,600 special drawing rights' are prospectively substituted for the words 'two hundred and fifty thousand francs' by the Carriage by Air and Road Act 1979, s 4 (1) (*a*) (i) (not yet in force).
[2] The words 'this limit' are prospectively substituted for the words 'two hundred and fifty thousand francs' by ibid., s 4 (1) (*a*) (ii) (not yet in force).
[3] The words '17 special drawing rights' are prospectively substituted for the words 'two hundred and fifty francs' by ibid., s 4 (1) (*a*) (i) (not yet in force).
[4] The words '332 special drawing rights' are prospectively substituted for the words 'five thousand francs' by ibid., s 4 (1) (*a*) (i) (not yet in force).
[5] For art. 22 (5) the following is prospectively substituted by ibid., s 4 (1) (*a*) (iii) (not yet in force):
'(5) The sums mentioned in terms of the special drawing right in this Article shall be deemed to refer to the special drawing right as defined by the International Monetary Fund. Conversion of the sums into national currencies shall, in case of judicial proceedings, be made according to the value of such currencies in terms of the special drawing right at the date of the judgment.'

ARTICLE 23

(1) Any provision tending to relieve the carrier of liability or to fix a lower limit than that which is laid down in this Convention shall be null and void, but the nullity of any such provision does not involve the nullity of the whole contract, which shall remain subject to the provisions of this Convention.

(2) Paragraph (1) of this Article shall not apply to provisions governing loss or damage resulting from the inherent defect, quality or vice of the cargo carried.

ARTICLE 24

(1) In the cases covered by Article 18 and 19 any action for damages, however founded, can only be brought subject to the conditions and limits set out in this Convention.

(2) In the cases covered by Article 17, the provisions of the preceding paragraph also apply, without prejudice to the questions as to who are the persons who have the right to bring suit and what are their respective rights.

ARTICLE 25

The limits of liability specified in Article 22 shall not apply if it is proved that the damage resulted from an act or omission of the carrier, his servants or agents, done with intent to cause damage or recklessly and with knowledge that damage would probably result; provided that, in the case of such act or omission of a servant or agent, it is also proved that he was acting within the scope of his employment.

ARTICLE 25A

(1) If an action is brought against a servant or agent of the carrier arising

out of damage to which this Convention relates, such servant or agent, if he proves that he acted within the scope of his employment, shall be entitled to avail himself of the limits of liability which that carrier himself is entitled to invoke under Article 22.

(2) The aggregate of the amounts recoverable from the carrier, his servants and agents, in that case, shall not exceed the said limits.

(3) The provisions of paragraphs (1) and (2) of this Article shall not apply if it is proved that the damage resulted from an act or omission of the servant or agent done with intent to cause damage or recklessly and with knowledge that damage would probably result.

ARTICLE 26

(1) Receipt by the person entitled to delivery of baggage or cargo without complaint is *prima facie* evidence that the same has been delivered in good condition and in accordance with the document of carriage.

(2) In the case of damage, the person entitled to delivery must complain to the carrier forthwith after the discovery of the damage, and, at the latest, within seven days from the date of receipt in the case of baggage and fourteen days from the date of receipt in the case of cargo. In the case of delay the complaint must be made at the latest within twenty-one days from the date on which the baggage or cargo have been placed at his disposal.

(3) Every complaint must be made in writing upon the document of carriage or by separate notice in writing despatched within the times aforesaid.

(4) Failing complaint within the times aforesaid, no action shall lie against the carrier save in the case of fraud on his part.

ARTICLE 27

In the case of the death of the person liable, an action for damages lies in accordance with the terms of this Convention against those legally representing his estate.

ARTICLE 28

(1) An action for damages must be brought, at the option of the plaintiff, in the territory of one of the High Contracting Parties, either before the court having jurisdiction where the carrier is ordinarily resident, or has his principal place of business, or has an establishment by which the contract has been made or before the court having jurisdiction at the place of destination.

(2) Questions of procedure shall be governed by the law of the court seised of the case.

ARTICLE 29

(1) The right to damages shall be extinguished if an action is not brought within two years reckoned from the date of arrival at the destination, or from the date on which the aircraft ought to have arrived, or from the date on which the carriage stopped.

(2) The method of calculating the period of limitation shall be determined by the law of the court seised of the case.

ARTICLE 30

(1) In the case of carriage to be performed by various successive carriers and falling within the definition set out in the third paragraph of Article 1, each carrier who accepts passengers, baggage or cargo is subjected to the rules set out in this Convention, and is deemed to be one of the contracting parties to the contract of carriage in so far as the contract deals with that part of the carriage which is performed under his supervision.

(2) In the case of carriage of this nature the passenger or his representative can take action only against the carrier who performed the carriage during which the accident or the delay occurred, save in the case where, by express agreement, the first carrier has assumed liability for the whole journey.

(3) As regards baggage or cargo, the passenger or consignor will have a right of action against the first carrier and the passenger or consignee who is entitled to delivery will have a right of action against the last carrier, and further, each may take action against the carrier who performed the carriage during which the destruction, loss, damage or delay took place. These carriers will be jointly and severally liable to the passenger or to the consignor or consignee.

CHAPTER IV   PROVISIONS RELATING TO COMBINED CARRIAGE

ARTICLE 31

(1) In the case of combined carriage performed partly by air and partly by any other mode of carriage, the provisions of this Convention apply only to the carriage by air provided that the carriage by air falls within the terms of Article 1.

(2) Nothing in this Convention shall prevent the parties in the case of combined carriage from inserting in the document of air carriage conditions relating to other modes of carriage provided that the provisions of this Convention are observed as regards the carriage by air.

CHAPTER V   GENERAL AND FINAL PROVISIONS

ARTICLE 32

Any clause contained in the contract and all special agreements entered into before the damage occurred by which the parties purport to infringe the rules laid down by this Convention, whether by deciding the law to be applied, or by altering the rules as to jurisdiction shall be null and void. Nevertheless for the carriage of cargo arbitration clauses are allowed subject to this Convention, if the arbitration is to take place within one of the jurisdictions referred to in the first paragraph of Article 28.

ARTICLE 33

Nothing contained in this Convention shall prevent the carrier either from refusing to enter into any contract of carriage, or from making regulations which do not conflict with the provisions of this Convention.

ARTICLE 34

The provisions of Articles 3 to 9 inclusive relating to documents of carriage shall not apply in the case of carriage performed in extraordinary circumstances outside the normal scope of an air carrier's business.

ARTICLE 35

The expression 'days' when used in this Convention means current days not working days.

ARTICLE 36

The Convention is drawn up in French in a single copy which shall remain deposited in the archives of the Ministry of Foreign Affairs of Poland and of which one duly certified copy shall be sent by the Polish Government to the Government of each of the High Contracting Parties.

ARTICLE 40

(1) [*This paragraph is not reprodured. It defines 'High Contracting Party'*[1]*.*]

(2) For the purposes of the Convention the word *territory* means not only the metropolitan territory of a State but also all other territories for the foreign relations of which that State is responsible.

[*Articles 37, 38, 39, 40 and 41 and the concluding words of the Convention are not reproduced. They deal with the coming into force of the Convention*[2].]

---

[1] Reproduced in full in appendix A, ante.

[2] Reproduced in full in appendix A, ante. The final clauses of the Hague Protocol are printed in that appendix.

ADDITIONAL PROTOCOL (*With reference to Article 2*)

The High Contracting Parties reserve to themselves the right to declare at the time of ratification or of accession that the first paragraph of Article 2 of this Convention shall not apply to international carriage by air performed directly by the State, its colonies, protectorates or mandated territories or by any other territory under its sovereignty, suzerainty or authority.

Bibliography

Legislation

Statutes

Carriage by Air Act, 1961.
Carriage by Air (Supplementary Provisions) Act, 1962.
Fatal Accidents Act, 1976.
Carriage by Air & Road Act, 1979.
Civil Aviation Act, 1982.
Civil Aviation (Eurocontrol) Act, 1983.
Airports Act, 1986.
Consumer Protection Act, 1987.
Civil Aviation (Air Navigation Charges) Act, 1989.
Environmental Protection Act, 1990.

Statutory instruments

Carriage by Air (Supplementary Provisions) Act, 1962 (Commencement) Order 1964, SI No. 486.
Carriage by Air Acts (Application of Provisions) Order 1967, SI No. 480.
Carriage by Air Acts (Application of Provisions) (Amendment) Order 1969, SI No. 1083.
Mortgaging of Aircraft, Order 1972 SI No. 1268.
Carriage by Air Acts (Application of Provisions) (Second Amendment) Order 1979, SI No. 931.
Mortgaging of Aircraft (Amendment) Order 1981 SI No. 611.
Carriage by Air Acts (Application of Provisions) (Third Amendment) Order 1981, SI No. 440.
Aircraft (Customs and Excise) Regulations 1981, SI No. 1259.
Civil Aviation (Eurocontrol) Act 1983 (Commencement No. 1) Order 1983, SI No. 1886.
Air Navigation (Aircraft & Aircraft Engine Emissions) Order 1986, SI No. 599.
Carriage by Air (Sterling Equivalents) Order 1986, SI No. 1778.
Air Navigation (Investigation of Air Accidents Involving Civil & Military Aircraft or Installations) Regulations 1986, SI No. 1953.
Carriage by Air (Parties To Convention) Order 1988, SI No. 243.

Air Navigation (Aeroplane & Aeroplane Engine Emission of Unburned Hydrocarbons) Order 1988, SI No. 1994.

Civil Aviation (Route Charges for Navigation Services) Regulations 1989, SI No. 303.

Air Navigation Order 1989, SI No. 2004.

Air Navigation (Amendment) Order 1990 SI No. 2154.

Civil Aviation (Investigation of Air Accidents) Regulations 1989, SI No. 2062.

Civil Aviation (Route Charges For Navigation Services) (Amendment) Regulations 1989, SI No. 2257.

Air Navigation (Noise Certification) Order 1990, SI No. 1514.

Civil Aviation (Navigation Services Charges) Regulations 1991, SI No. 470.

Civil Aviation Authority Regulations 1991, SI No. 1672.

Cases

Donoghue V Stevenson [1932] A.C. 562.

Grein V Imperial Airways Limited [1937] 1 KB, (CA).

Collins V British Airways Board (1981) CA.

Petrire V Spantax S.A. (1985) US Ct. of Apps., 2nd Circ.

Swiss Bank Corporation V Brinks M.A.T. Limited (1986) QBD.

Goldman V Thai Airways International Limited (1983) CA.

Trans World Airlines Inc. V Franklin Mint Corporation (1984) (US SC).

Customs and Excise Comrs. V. Air Canada [1990] CA.

Civil Aviation Authority

Britain's Civil Aviation Authority, C.A.A. Document No. 225 (1986).

Review of the Civil Aviation Authority 1984, C.A.P. 486.

Deregulation of Air Transport: A Perspective on the Experience in the United States, 1984, C.A.A. PAPER 84009.

Airline Competition Policy, 1984, C.A.P. 500.

Statement Of Policies On Air Transport Licensing, June 1988, C.A.P. 539.

The Common Market

Bulletin of the European Communities: Supplement 5/79 *Air Transport: a Community Approach Memorandum of the Commission*, 1979.

Council Decision 80/50/EEC.

Council Directive 80/51/EEC.

Council Directive 80/1266/EEC.

Commission of the European Communities: Memorandum No. 2. *Progress towards the development of a Community Air Transport Policy*. Com (84) 72 final: Brussels, 15 March 1984,

Council Regulation (EEC) No. 3975/87.

Council Regulation (EEC) No. 3976/87.

Commission Regulation (EEC) No. 4261/88.

Council Regulation (EEC) No. 2299/89.

Council Directive 89/629/EEC.
Council Regulation (EEC) No. 2342/90.
Council Regulation (EEC) No. 2343/90.
Council Regulation (EEC) No. 2344/90.
Commission Regulation (EEC) No. 82/91.
Commission Regulation (EEC) No. 83/91.
Commission Regulation (EEC) No. 84/91.
Council Regulation (EEC) No. 294/91.
Council Regulation (EEC) No. 295/91.

International Treaties

Warsaw Convention, 1929.
Chicago Convention, 1944
    International Air Services ('Two Freedoms') Agreement
    International Air Transport ('Five Freedoms') Agreement.
Hague Protocol, 1955.
Brussels Convention, 1960.
Guadalajara Convention, 1961.
Guatemala Protocol, 1971.
Montreal Protocols, 1975.
Brussels Protocol, 1981.

International Civil Aviation Organisation

International Standards and Recommended Practices, Environmental Protection: Annex 16, Volume I – Aircraft Noise.
International Standards and Recommended Practices, Environmental Protection: Annex 16, Volume II – Aircraft Engine Emissions.

Miscellaneous

British Air Transport in the Seventies (Report of the Committee of Inquiry into Civil Air Transport), 1969, Cmnd. 4018.
Free Trade in the Air: Report of the Think Tank on Multilateral Aviation Liberalisation, Global Aviation Associates Limited, Jan. 1991.
Airlines, Tourism and the Environment, Stephen Wheatcroft, in Tourism Management, June 1991.
Air Law, Peter Martin, David McClean, Elizabeth Martin, Rod Margo, 4th Edition published by Shawcross & Beaumont.
Flying off Course — The Economics of International Airlines by Rigas Doganis published by Harper Collins.

Index